THE U.S.
IN SPACE

THE U.S. IN SPACE

Issues and Policy Choices for a New Era

EDMUND S. MUSKIE
Editor

CENTER FOR NATIONAL POLICY PRESS

WASHINGTON, D.C.

Copyright © 1988 by

Center for National Policy Press

Distributed by arrangement with

National Book Network
4720 Boston Way
Lanham, MD 20706

European Distribution by

Eurospan
3 Henrietta Street
London, WC2E 8LU England

Printed in the United States of America

British Cataloging in Publication Information Available

Library of Congress Cataloging-in-Publication Data

The U.S. in space : issues and policy choices for a new era / Edmund S. Muskie, editor.
p.   cm.
Includes bibliographies.
1. Astronautics and state—United States. I. Muskie, Edmund S., 1914–
II. Title: US in space.
TL789.8.U5 1988   629.4'0973—dc19   87–30956 CIP
ISBN 0–944237–23–1 (alk. paper)
ISBN 0–944237–24–X (pbk. : alk. paper)

CONTENTS

INTRODUCTION

SPECIALISTS IN THE field of space policy, including the authors in this volume, generally seem to agree that we are at a critical point for the U.S. space program. Major decisions, about scope, scale and direction, must be made again, much as they were in the late 1950s and early 1960s. This time, however, the decision-making landscape is vastly more complex than it was before, and we no longer have the convenience of addressing U.S. purposes and resources independently of the purposes and resources of other nations.

The issues and policy choices described in this publication include the problem of how best to protect a highly developed and essential system of military satellites now that technological advances in both the U.S. and the Soviet Union make attack conceivable; how to manage public and private investment in key space transportation programs; how to balance and take advantage of both competitive and cooperative forces in a complex international space marketplace; and, what kind of major basic research projects to mount for the future.

None of these issues existed in their current form or intensity when the U.S. space program first emerged into the public eye some 30 years ago.

In almost all respects, that program has represented a set of achievements characteristic of this nation's particular strengths and special vision.

As did enactment of the Marshall Plan, and before that the Morrill Act, the formulation of U.S. space policy broke the bonds of existing concepts and laid down a new framework for major structural developments of the future. At the core of this framework, establishing the concept of earth orbits as a set of internationalized "public ways," analogous to the major oceans of the planet, was critical to maintaining the possibility of an open, accessible, and fundamentally peaceful environment in near space. It paved the way for making regulated, rather than hostile competition for orbital territory the norm, and was consistent with the effort to constrain development of space-based weapons.

U.S. international leadership in this critical area was matched by a set of creative decisions about our own internal program that permitted significant progress on a number of different objectives.

Born initially out of advances in military technology generated by World War II, the space program ultimately captured the public imagination primarily with civilian activities. Over time, civilian and military purposes have managed to co-exist within a set of organizational structures deliberately designed to separate without creating conflicting objectives.

The U.S. space program has been, too, throughout its history a fairly harmonious joint effort of the public and private sectors. Private businesses from numerous major industries have operated effectively within the framework of programs organized and operated by the Department of Defense and NASA. And, until recently, the question of how much in the way of national resources to devote to basic scientific missions in space as opposed to more applied activities was resolved without exceptional difficulty.

Suddenly, however, what for years we took for granted as a highly successful exercise in national policy-making seems to be everywhere falling into clouds of doubt, confusion and even outright failure. How has this come about?

Together, the Soviet launch of Sputnik in 1957 and President Kennedy's dramatic announcement of the Moon program in 1961 focused the American public's attention on space and provided the impetus for significant investment of public resources in space activities. After the excitement of the early Apollo missions, however, the pressure of more prosaic and programmatic concerns—including the war in Vietnam, the war on poverty, a stalled domestic economy and a threatened domestic environment—drew attention and commitment away from NASA's activities. As was probably inevitable after the zeal that infused NASA during the Apollo effort, a more normal, multi-objective, less dramatic program drew a more modest level of investment.

In spite of its decline in national visibility, however, the U.S. program for development of the uses of space was significant and highly successful throughout the 1960s and 1970s, with an active emphasis on applications, and some major scientific expeditions into outer space, leading the world's advance into this ultimate human frontier.

The U.S. led the way in developing satellite systems for two extremely important applications, communications and remote sensing. We were also responsible for promoting the concept that space should be accessible for private as well as public sector activities. Finally, we promoted and encouraged the interests of the world scientific community in conducting basic scientific observations in and from space. None of these extremely important achievements of U.S. space policy, however, has been the product of the kind of focused, highly-motivated national effort that the original Moon program represented. As a result, under the pressures and difficulties that naturally arise in any complicated organizational effort, without a strong consensus about objectives and what they are worth, the sense of mission in our civilian space activities has been difficult to maintain.

Beginning in 1983, President Reagan's advocacy of the Strategic Defense Initiative, or SDI, did generate an immediate and heated debate about the militarization of space, and brought public attention sharply into focus on space policy after a hiatus of some years. But this strong military thrust in the public discussion about space came at a time when the civilian space program needed to be re-energized and perhaps redirected.

In a political atmosphere that had come close to indifference, where it was not locked in controversy, the Challenger disaster of January 1986 was a major setback for the U.S. space program as well as a significant human tragedy. Unlike earlier space program failures, this one caught us without a clear sense of continuing purpose and objective. Subsequent launch failures added to the confusion.

The shuttle program had to an extent recaptured the public imagination, but it was the end product of a series of earlier programmatic decisions, not a step along a new road. The issues of how much to invest in shuttle technology, how much to rely instead on expendable launch vehicles, and how to balance public purpose and private interest were questions that had concerned specialists for some time, but they had largely escaped the attention of the general public. Similarly, the possibility that the U.S. might be at a competitive disadvantage in space launch capacity *vis-á-vis* an emerging nation such as China, as well as Europe and other countries with advanced high-technology sectors, had been simply beyond public imagination. But as Edwin Deagle and Marcia Smith point out in their articles, there is a complex mix of public and private entities that now comprise the world space industry, and U.S. space policy must be formed in awareness of how that entire industry—domestic and international—is likely to develop.

Even in the early days of U.S.-Soviet competition, there was room to configure U.S. policy in a way that avoided impossibly difficult trade-offs. The new world of space was open for exploration and development, and as a major nation, with extensive resources, talent, and energy, we had the luxury of creating a plan for entering this new world that allowed for a great deal of freedom in selecting objectives, and setting whatever pace we felt necessary to achieve them. Our task now, however, is to move forward effectively in an era of constraints, and in an environment of strong competition, not only from the Soviet Union but also from many other nations.

ADDRESSING THE MAJOR ISSUES

As noted, the authors in this volume address several critical issues for the future of U.S. policy: how best to ensure the security of vital military and civilian satellites; how best to ensure a U.S. competitive edge in space transportation technology; how to secure the U.S. a major place in the future of space by balancing competitive and cooperative strategies in dealing with other nations; and how we might benefit from investments in major scientific research in space.

There are a number of trade-offs that we will have to confront as a nation as we address these issues.

One, of course, is how far to go with the SDI.

Since its inception in 1983, the Congress has had to decide each year on a level of resources to commit to this ambitious and controversial program.

Thus far, even in a period in which efforts to control the federal budget deficit have been a major, overriding concern, Congress has authorized and appropriated substantial sums for basic SDI research. While the work that comes under the umbrella of SDI relates in the first instance to national security concerns, and not to space policy *per se*, it nonetheless draws on a body of research and technology, and on human and material resources, that are part of our national space program infrastructure. The work that will be done in the service of SDI is classified and militarily oriented in specification, and much of it will be diffused into civilian activities only after a considerable time lag, if at all.

At the same time, as Steve Fetter documents in his article on the protection of current U.S. space assets, an active SDI program would place in jeopardy the possibility of continuing restraints on weaponry in space.

Were an SDI system to be deployed, it would represent abandonment of fundamental policies about the uses of space that the United

States successfully promoted and negotiated in the early years of U.S.-Soviet competition. In the late 1950s and early 1960s, as the United States and the Soviet government were assessing the nature of the threat posed by each other's advances in space technology, decisions were made that were to determine the nature of military activities in space for two decades. These decisions focused on the effort to confine all military space assets to essentially non-aggressive devices oriented to reconnaissance and communications, to prohibit the orbiting of weapons, and to confine manned space missions to non-military purposes.

From the U.S. perspective, this was the prudent way to ensure that our national security would not be seriously undermined, since we perceived the military usefulness of space-based reconnaissance and communications, in which we had the likely technological advantage, and we feared the possibility that the Soviets might develop a nuclear weapon that would be placed in orbit.

The existence of this tacit arrangement, which was called an agreement to conduct only "peaceful" activities in space, has in fact enabled us to place greater and greater dependence on space systems for normal military operations (e.g., submarine navigation and communications). The continuation of this regime is fundamentally contradicted by the SDI concept.

The SDI program is not a major focus of the articles in this volume, since the authors are dealing with basic issues of space policy that would be of concern whether or not the SDI proposal had been made. The level and degree of commitment to SDI that the U.S. eventually might make, however, will directly affect the resolution of many if not all of the rest of these other issues of space policy.

It is important, for this reason, that as Congress and the public review the SDI program, attention is given to its interaction with the broad array of policy questions laid out by these authors, including the protection of existing space assets, the investment of resources in space science and space technology, and the direction of both our competitive and cooperative participation in the international space community.

At the other end of the spectrum of human concerns from the possibility of nuclear annihilation on earth is the possibility of discovering, or of placing, life in the far reaches of the universe.

Scientists have been expressing increasing concern over the focus and direction of U.S. space policy. The immediate issues are primarily centered on the constraints on scientific launches posed by our present emphasis on the shuttle, with its continuing delays and problems. But not far in the background is the central question of a new U.S. space agenda.

At the outer boundaries of known possibilities lie observations of the far reaches of the universe such as those set forth by Frank Drake in his article in this volume. Herein lies a research agenda for space that might, finally, put us on the edge of discovery of other life forms, or at least of the chemical and physical conditions that might create and sustain them, an agenda for broadening and deepening our understanding of the basic forces that comprise the universe and its components.

The National Commission on Space has recommended a long-range civilian space program that includes basic scientific advances, an ambitious and creative program for Solar System exploration, and further investment in practical applications of space technology.

Whatever specific decisions we take over the next several years, the resources and commitment necessary to carry us forward into the next Space Age will strain our budgets, if not our imaginations. U.S. Senator Spark M. Matsunaga and others have proposed that we cooperate actively with the Soviet Union in order to accomplish major space objectives, such as exploration of Mars. Not all would agree that such an effort is desirable, but, as Marcia Smith makes clear, great accomplishments on the space frontier are already being made by other nations, and the U.S. itself now has much to gain from international cooperation in space.

While the means and the ways of space policy will remain subject to argument for some years to come, it is clear that there is a need for vision, a vision which ought to inspire all of us, whatever our age and whatever else we might ask of ourselves and our society, to make the commitments necessary to reach beyond the difficulties, the trade-offs, and the constraints—in short, to reach beyond imagination, beyond the bounds of earth, to both the planets and the stars.

—Edmund S. Muskie

PROTECTING OUR MILITARY SPACE SYSTEMS

Steve Fetter

OVER THE LAST 25 years the United States has become increasingly dependent on space-based systems to support its military forces, and this trend is likely to continue for some time. Satellite systems have become an integral part of nuclear deterrence by providing strategic warning of an attack, tactical warning of missile launches, reliable communications between command authorities and nuclear forces, and nuclear explosion detection. Satellites also aid in conventional warfighting by providing accurate reconnaissance, intelligence, weather, and navigation information.

Current and future anti-satellite (ASAT) weapon technologies are capable of preventing many of our space systems from carrying out their missions, thereby possibly decreasing the stability of nuclear deterrence and weakening the effectiveness of conventional forces. There is a broad range of policy options available that, in principle, could help to protect our space assets. While unilateral measures could go a long way toward safeguarding satellite systems, a review of these options makes it clear that bilateral agreements are also necessary if we are to guard against the full range of ASAT threats without generating dangerous instabilities.

WHICH SATELLITES TO PROTECT?

Not all satellites are equally important to our national security. The United States currently performs four basic types of military missions with satellites that are of interest here:[1]

Communications. About a dozen satellites, grouped in four satellite systems, are used for military and diplomatic communications. Except for two satellites that relay messages to the polar regions, all U.S. military communications satellites are in geostationary orbit (GSO)

1

36,000 km above the surface of the earth.[2] In addition to the systems already in use, an advanced, interservice satellite communications system called MILSTAR (Military Strategic and Tactical Relay) is now under development, with deployment planned for the early 1990s. The MILSTAR system, which will consist of a half-dozen satellites in inclined geosynchronous orbits, is intended to provide command and control communications at all levels of conflict, including general nuclear war.

Navigation. The U.S. has two military navigation satellite systems: Transit and NAVSTAR. Transit, which travels in low earth orbit (LEO) at an altitude of about 1,000 km, was developed to aid in the navigation of Polaris submarines. The much newer NAVSTAR system, when complete, will consist of 18 satellites 20,000 km above the earth. Radio signals emitted from the navigation satellites can be used by special receivers on the earth to obtain very accurate position and velocity information.

Meteorology. Two DMSP (Defense Meteorological Satellite Program) satellites in LEO process visible and infrared images of the earth to provide information on cloud cover, temperature, and precipitation world-wide.

Reconnaissance and surveillance. Under this broad category are grouped several systems that observe electromagnetic signals reflected or emitted from objects on earth. These systems serve different missions: attack warning, nuclear burst detection, photoreconnaissance, and electronic surveillance. Attack warning is provided by three DSP (Defense Support Program) satellites in GSO that detect the infrared emissions of missiles as they are launched. Sensors on two dozen satellites, including those in the NAVSTAR system, can detect and locate nuclear explosions. Photoreconnaissance and electronic surveillance are highly classified programs, but it can be said that a small number of photoreconnaissance satellites travel in LEO, sometimes at altitudes less than 200 km, to obtain high-resolution photographs for use in treaty verification and intelligence.

The Soviet Union uses satellites for the same missions, but there are three important differences between U.S. and Soviet systems: (a) Soviet satellites have shorter lifetimes, (b) the U.S.S.R. has more single-purpose satellites, and (c) the Soviets have a large number of satellites in Molniya (highly-elliptical) orbits instead of in GSO. The first two factors combine to give the U.S.S.R. a launch rate five times greater[3] and a total constellation size nearly twice as great[4] as the U.S. This does not mean, however, that the U.S.S.R. has an advantage over the U.S. in space capability. Although it may be true that the Soviets can reconstitute satellite systems more quickly in the event of their destruction by ASATs, it is not clear how valuable this would be in an actual conflict (see below). The third factor may make it more difficult for the Soviets to protect their satellites, because satellites in Molniya orbits pass much closer to the earth than satellites in GSO, and therefore are much more vulnerable to ASATs based on earth.

Which satellite systems need protection most? This question is most easily answered when considering systems that are vital to nuclear deterrence, since increasing the stability of deterrence is clearly in the interests of both sides. To the extent that deterrence depends on both sides having tactical warning of an attack, it is especially important that both sides have confidence in the security of their attack warning satellites. Although attack warning is also provided by ground-based radars, satellites can detect missile launches 15 minutes earlier, doubling the time available for decisions and relaxing the need for a hair-trigger response. More important, without attack warning satellites we would have to rely solely upon radars—there would be no independent confirmation that an attack was underway.

The ability to communicate orders to surviving nuclear forces is also essential to deterrence. At present, the weak link in deterrence is not so much the likelihood that sufficient nuclear forces would survive, but rather the likelihood that the ability to command them would remain after a first strike. To the extent that the ability to maintain continuous communication between the National Command Authority and strategic and tactical forces depends on communication satellites (and the geographically dispersed nature of Western forces virtually requires their use), safeguarding these satellites is essential.

Turning to other military space missions, two appear important, but not essential to deterrence. Navigation satellites successfully targeted during a nuclear war would deny U.S. bombers and submarines accurate guidance information for the destruction of hard, or especially well-defended, targets, but this information would not be required in order to destroy most military, industrial, and population centers. Satellites capable of detecting nuclear bursts may also be tempting targets, since they could be used to assess the success of a U.S. strike or the damage from a Soviet strike. But navigation and nuclear burst detection are clearly less vital missions in the maintenance of deterrence.

Meteorological satellites, while very valuable in peacetime and during conventional wars, are less important to nuclear deterrence. Photoreconnaissance satellites are similarly valuable in peacetime, to monitor compliance with arms control treaties, and during conventional wars and crises. But they may become threatening during nuclear war since they could locate surviving forces for retargeting. If necessary, their missions could be performed adequately by aircraft or fractional-orbit satellites in wartime.

Of the missions performed by satellites now in orbit, attack warning and communications are the most essential for maintaining confidence in nuclear deterrence. We should on that count endeavor to ensure their survival, perhaps even if, to do so, we must accept agreements that help ensure the survival of the same functions for the other side. Safeguarding these systems should be in the interests of both sides, since increasing confidence in their survivability increases the crisis stability of nuclear deterrent forces, thus making preemptive or inadvertent war less likely.

With respect to conventional war, it is much more difficult to determine which satellites should be protected. Attack warning, strategic communications, and nuclear burst detection are irrelevant (unless one is planning to escalate the conflict to the nuclear level). On the other hand, tactical communications, navigation, meteorological, and reconnaissance satellites can aid both sides substantially in targeting enemy forces. The latter systems are force multipliers, and each side will naturally seek to preserve its own capabilities while denying such capabilities to the other side.

The following discussion focuses primarily on protecting U.S. satellites critical to nuclear deterrence, such as those used for attack warning and communications. These are the systems for which bilateral agreements are most likely to succeed. It should be noted, however, that technology developed to attack other systems could threaten these critical satellites, although the high orbits of attack warning and communications satellites serve to make attack on them much more difficult, time consuming, and costly.

THE ASAT THREAT

In the broadest sense, an anti-satellite (ASAT) weapon system is any type of weapon system that can be used to interfere with the mission of a satellite. This includes not only damaging or destroying satellites, but also jamming communications and destroying ground facilities. Here we discuss only satellite destruction, however, because the latest communication satellites (e.g., MILSTAR) are virtually jam-proof, [5] and because ground stations can be made much less vulnerable than satellites through proliferation and mobility.

Earth-based ASATs

All of the current weapon systems that have potential or inherent ASAT capability are based on earth. Earth-based ASATs have the primary advantages of being larger, less vulnerable, and much cheaper to construct and maintain than ASATs based in space. They have the disadvantages of being far from targets in GSO, and of having to cope with the limitations imposed by the earth's atmosphere. The two primary types of ASATs are missiles and directed-energy weapons.

Missiles

Ground, sea, or air-launched missiles can be used to attack satellites provided that their range is sufficient to reach the satellite orbit in question. This includes not only missiles intended for ASAT use, such as the U.S. air-launched ASAT now in development or the Soviet ground-launched ASAT (both of which are only capable of attacking satellites in LEO), but also nuclear-armed intercontinental and sub-marine-launched ballistic missiles (ICBMs and SLBMs) and the Sprint and Galosh anti-ballistic missiles (ABMs).

With modifications that could be developed and tested in a few years, such as a lighter payload, proper fusing, or an additional rocket stage, ICBMS, SLBMs, ABMs and current ASATs could destroy satellites in GSO. These weapons could use nuclear warheads of various yields or conventional homing warheads. ASATs using nuclear warheads have a damage radius ranging from tens to thousands of kilometers, depending on the yield of the weapon and the hardness of the target satellite, but they could also damage unhardened friendly satellites. Conventional warheads would have to come much closer to the target—within at least one kilometer—in order to be effective.

The speed of such missiles is on the order of ten kilometers per second, so it would take at least an hour to reach attack-warning and communications satellites in GSO. In the case of conventional warheads, mid-course update and terminal homing guidance would be required for adequate accuracy. More sophisticated space tracking systems than those currently in use would probably also be required. Any attack on satellites in GSO with earth-based missiles would be detectable by attack-warning satellites, since the rocket boosters required to reach high altitudes are very large. There would be sufficient time to discover the purpose of the missile and alert the nuclear forces, hence fulfilling at least some of the attack warning and communications missions of the satellites. Because they are slow and detectable, earth-based missiles probably do not represent the most dangerous threat to satellites in GSO.

The situation for satellites in LEO is quite different. Earth-based missiles can reach satellites orbiting 200 to 1000 km above the earth in a few minutes, and guidance technologies have already been proven effective enough to use conventional warheads (or no warhead at all, by striking the satellite directly) at these distances.

Directed-energy weapons

Ground-based high-energy lasers (HELs) of certain wavelengths could destroy satellites through heating or shock. HELs have the advantage of delivering energy fast—only a tenth of a second is needed to reach GSO from the ground—and the disadvantages of being large and inefficient in terms of the amount of energy required to destroy a target. Several types of HELs (chemical, free-electron, and excimer lasers) under development are potential earth-based ASAT weapons.

Much has been said about the destructive potential of HELs against space objects, but it is important to note the differing requirements for damaging one unhardened satellite under test conditions, and attacking hardened satellite systems during war. Although a high-quality one megawatt ground-based laser, which may be feasible in the next few years, could damage or destroy unhardened satellites in GSO by irradiating them for tens of minutes, this does not mean that effective ASATs could be based on such lasers. An actual attack against a hardened communications satellite system, for example, would require the certain destruction of several satellites in less than a minute. A laser weapon capable of this task would emit at least a hundred megawatts of power. It is likely to take some time to develop such large lasers, and even then it may prove impossible to transmit such large amounts of power through the atmosphere. At least five such lasers located around the world would be needed to provide continuous coverage of all GSO satellite targets.

As noted above, however, satellites in LEO are much less demanding targets. Most could be destroyed by an ASAT system based on the current state-of-the-art chemical laser, although at large cost. Even very hard satellites could be destroyed with lasers that will probably become available in the 1990s.

Space-based ASATs

Weapons based in space can be much closer to and have a clearer view of their targets than earth-based ASATs; hence, for a given level of technology, they can be more effective against satellites in any orbit. Space-based ASATs are more vulnerable, however, and the costs associated with deploying and maintaining ASATs in space are much greater than the costs of earth-based ASATs. Three types of potential space-based ASAT weapons are explored here: space mines, kinetic-energy weapons, and directed-energy weapons.

Space mines

A space mine would be a satellite that is placed in the same orbit as a target satellite (usually well in advance of an attack), and which attempts to remain within lethal range at all times. Space mines could be salvage-fused, meaning that any attempt to interfere with them would cause them to explode, destroying the target satellite in the process. Although no space mines are known to exist, they could probably be developed in a few years.

If armed with conventional warheads, a space mine would have to stay within a kilometer of the target at all times, a requirement that may be difficult to meet. Nuclear space mines would be destructive at much larger distances (tens or hundreds of kilometers), but they could also destroy or interfere with friendly satellites nearby. The deployment of nuclear weapons in space is banned by the Outer Space Treaty. The deployment of space mines during peacetime could pose the danger of inadvertent war through accidental detonation, and their use during a conventional war to destroy reconnaissance satellites could contribute to the likelihood of escalation.

Kinetic-energy weapons

Kinetic-energy weapons, either homing missiles or projectiles fired from guns, could be used to destroy satellites by direct impact. Rail guns, which use powerful magnetic fields to accelerate projectiles, are theoretically capable of much higher velocities than are achievable with rockets, although current devices are far less capable than rockets. A rail gun would weigh hundreds of tons and would not be cost-effective unless it could destroy many satellites.[6]

Homing missiles could attack satellites in GSO in several minutes if they were "parked" in orbits a few hundred kilometers from the target satellite. (This concept differs from a space mine in that the target normally remains outside lethal range.) The size and technological requirements of such a missile would be comparable to those of the current U.S. air-launched ASAT homing missile, and therefore are feasible in the near term. Such missiles could carry low-yield nuclear weapons, which would considerably reduce the tracking requirements and be more robust to defensive countermeasures, but would have the liabilities of nuclear use noted above.

Space-based directed-energy weapons

Directed-energy weapons based in space can use uncharged particles (photons or neutral atoms) of any energy. (Charged particle beams cannot be used as ASAT weapons because the earth's magnetic field deflects them.) Candidate technologies, none of which are currently judged feasible in the short term, are neutral particle beams, the high-energy lasers discussed above, X-ray lasers, and microwave weapons.

Neutral particle beams. Neutral particle beam (NPB) weapons, which are the best-developed directed-energy weapons suitable for space deployment, are similar to the accelerators used by particle physicists. The particle energy is limited by the size of the accelerator; current design concepts generate particles with energies of a few hundred million electron volts (MeV). Protons of this energy have the capacity to penetrate to the center of a target satellite and destroy or damage its electronics. It may be possible to harden electronics, which would lead to a corresponding increase in the amount of time the NPB would have to irradiate the target. Even so, NPBs, if they can be constructed in space, might be very effective against satellites in GSO. Since a space-based NPB weapon is unlikely to cost less than a target satellite, its range would have to be large enough to engage several targets at once. An alternate approach would be to put an NPB in a highly-elliptical orbit that intersects GSO or into counter-rotating GSO, allowing a single low-power weapon to attack all satellites in GSO, though over an extended period of time (at least 12 hours). The power requirements for an NPB could be quite large. Depending on the circumstances, over 10 tons of fuel might be required to destroy a 1-ton satellite. An NPB would have a linear dimension of perhaps 50 meters, making it a very noticeable object.[7]

Space-based HELs. HELs based in space could be much smaller than earth-based HELs because they could be placed close to target satellites, but as the range of the laser decreases so does the number of targets that can be attacked in a given amount of time. For example, less than one-thousandth the power of an earth-based laser would be required to destroy a target in GSO from a distance of 1,000 km, but then the laser could only attack a single satellite at a time. It is very difficult to hide a space-based HEL, because reducing the size of the mirror (which might be several meters in diameter) increases the power requirements. Even a low-power space-based laser would be quite noticeable and identifiable.

X-ray lasers. X-ray lasers have the primary advantage of a very compact and extremely high-power energy source: a nuclear explosion. This allows the possibility that X-ray lasers based in space might not be identifiable, or that they could be put on earth-based missiles and fired as soon as they were above the atmosphere (within a few minutes). They would require the launching and exploding of nuclear weapons in space. In theory, they could be very effective ASATs, capable of destroying instantaneously a number of satellites at very long ranges. They are in the research stage and presently it is not possible to say what can be attained, however.

Microwave weapons. It may be possible to build a device that uses a nuclear explosion to generate a narrow beam of microwaves. Electronic circuits are probably at least three orders of magnitude more vulnerable to microwave energy than to X-ray or particle-beam energy. On the other hand, the destruction of electronics would not usually be noticeable from the outside of a satellite, leading to uncertainty about whether the target had been disabled. Much more work needs to be done on this concept before further judgments can be made about its ultimate usefulness as a weapon.

In the preceding discussion, weapon system size requirements have been posited in terms of their effectiveness relative to U.S. estimates of U.S. satellite vulnerability. Actual ASAT weapons would likely have to be much more powerful (by perhaps an order of magnitude). For any given destructive mechanism there is a fairly well-defined threshold beyond which a given satellite will fail. The defender, who knows the details of the satellite design, can estimate this threshold with some confidence, although there will always be some uncertainty in the estimate of the system's vulnerability. Prudence requires that the estimate be conservative, so one can be confident that the system could survive an attack of that magnitude. The attacker, on the other hand, does not know the details of the design, and will be inclined to substantially overestimate the lethal level required to be confident that the target system will be destroyed. In most cases an attacker will want a "hard kill" of the system (i.e., the damage to target must be easily visible), which also increases the size requirement of the weapon.

POSSIBLE COUNTERMEASURES AGAINST ASATs

Unilateral countermeasures

Unilateral countermeasures are actions that the United States or the Soviet Union could take to safeguard satellites without the cooperation of other countries. Since the ASAT threat to high-altitude satellite systems is as yet undefined, actions taken should be those effective against a wide range of technologies.

Passive countermeasures

Passive unilateral countermeasures are those that enable the satellite to withstand or avoid the attack by ASATs. These may include (a) hardening the satellite against attack mechanisms (heating, shock, irradiation, and jamming); (b) evasion (maneuvering, hiding, and use of decoys); (c) redundancy (spares in orbit or ready to launch, and land-based back-ups to space-based systems); and, (d) placing satellites in less vulnerable orbits.

Hardening can be achieved by making the working components of the satellite (e.g., solar cells or microprocessors) less vulnerable to the ASAT threat and/or by surrounding the satellite or vulnerable components with an appropriate shield. Examples of internal hardening are radiation-resistant electronics to protect against the effects of nuclear weapons or NPBs, or heat-resistant components to withstand laser heating. Examples of shields are multicomponent X-ray shields against nuclear weapons and reflective or disintegrative shields and sensor shutters against lasers. The measures taken to address one class of threat must be consistent with and complementary to those taken to guard against other threats.

These hardening measures can go a long way toward reducing satellite vulnerability, and can also have a favorable cost-exchange ratio against the offense. For example, hardening electronics to levels feasible in the near term would force a long-range NPB weapon to consume an amount of fuel much more massive than the satellite it is attacking. The attacker might compensate by bringing the NPB closer to its targets, but this would require the construction of additional (expensive) NPB weapons. Another example is hardening against continuous wave (CW) lasers. Measures that increase the hardening of satellites by a factor of 10 might cost about 10 percent of total satellite costs,[8] but would require a laser 10 times more powerful— and much more costly—to destroy those satellites.

Cost-effectiveness trade-offs are different with regard to hardening against nuclear weapons or kinetic-energy weapons. The cost of a nuclear weapon is not proportional to its yield: for example, a 100 kiloton (kt) weapon may not cost much more than a 10 kt weapon.[9] Hardening against nuclear weapons can prevent the destruction of more than one satellite by a single weapon, or, in the case of an X-ray laser, decrease the number of satellites a single X-ray laser can destroy. For a given yield, hardening can also force the attacker to come closer to the satellite, thereby increasing warning time and the opportunity for maneuvering.

However, neither a nuclear explosion dedicated to the destruction of a single satellite nor, in the case of kinetic-energy weapons, a direct hit by a projectile, can be countered by any reasonable level of hardening. It is also very difficult to harden against the shock effects caused by very short-duration ("pulsed") chemical or X-ray lasers. For cost-effective unilateral countermeasures against nuclear, kinetic-energy, or pulsed-laser weapons, one must turn to the tactics of evasion and proliferation.

Maneuvering is effective against some threats. High-orbit satellites, for example, need only modest maneuvering capability to evade nuclear-armed earth-based missiles. This is true even if the missiles have terminal homing guidance, because the homing system is not effective until the missile is within a few dozen kilometers of the target, and the satellite can escape from this small volume of space in which the missile expects to find its target before the homing system is turned on. If the ASAT missile is space-based less than a thousand kilometers from the satellite, however, the fuel requirements for effective maneuvering would be excessive, because the satellite would have to accelerate very quickly to escape the homing system. Maneuvering can be defeated by giving missiles mid-course update capability, so they can track the maneuvers of target satellites during the entire missile flight. Maneuvering doesn't help at all if there is no warning, as would be the case with directed-energy weapons or close-by space mines.

Decoys may be deployed before or after attack. If decoys are deployed before an attack, the attacker can examine them. The decoys therefore must be realistic and, consequently, expensive. This cost could be reduced by keeping a number of inactive satellites in orbit, and having the decoys mimic these. This is sometimes called "anti-simulation," because one is trying to make the real thing (the inactive satellites) look like the cheap decoys, rather than trying to make expensive decoys that mimic the functions of active satellites. But even inactive satellites must perform a number of sophisticated functions, especially if the military is to trust that these systems would perform properly when called upon. Testing the inactive satellites would, however, let the other side know that they were real and not decoys. If warning of an ASAT attack is available, cheap decoys could be deployed at the moment of attack, but this strategy will only work against non-nuclear homing missiles—ballistic missiles are not smart enough to be fooled by decoys (they have to be targeted on something that is already there), and directed-energy weapons give no warning.

Another form of evasion is hiding, which might be effective against ground-based radar and optical satellite space tracking systems. But, even though the current U.S. optical tracking system can detect nothing smaller than a one meter-sized object at geosynchronous distances,[10] nearly all satellites require exposed components (antennae, solar cells, etc.) that are difficult to conceal. Space-based satellite surveillance systems that use a larger part of the electromagnetic spectrum could make hiding near the earth all but impossible (e.g., reducing the optical signal of a satellite by painting it black inevitably increases its infrared signal by making it hotter).

Redundancy is another possibility. If one could increase the number of hardened satellites by a substantial factor, one would, at the very least, force the ASAT system to become large and obvious. But redundancy alone tends not to be not cost-effective, because if an ASAT system is cost-effective against some number of satellites, it is likely to be cost-effective against twice that number. The advantage might go to the defense, however, if it were possible to replace a few complex, expensive, multi-purpose satellites with many simpler, cheaper, single-purpose satellites, but this is certainly not the current trend in U.S. satellite design. Soviet satellite systems tend to be more redundant than their U.S. counterparts.

A related countermeasure is reconstitution, which is the ability to quickly replace satellites that have been destroyed. This is not a practical option for the satellites and time scales that we are most concerned about. Even if replacement satellites were stockpiled (at a cost of several hundred million dollars each), they would take many hours to launch, and many more hours to maneuver into orbit.

Another potentially effective passive countermeasure would be to move critical satellites into higher, less-crowded orbits. These non-GSO orbits offer several advantages: (a) ground-based ASAT missiles would take longer to reach higher altitudes, increasing the time available for warning, maneuvering, and decoy deployment; (b) the power requirement for a ground-based laser ASAT is proportional to the square of the satellite altitude, so that a target at 10 times GSO would require laser power 100 times greater to destroy than the same target would require in GSO; (c) there are many non-synchronous orbits, in contrast to the single GSO, so such orbits would be less crowded and the identification of potentially hostile satellites or space mines much easier; and (d) satellites in higher orbits are more difficult to track from the ground, which could frustrate ASAT attacks.

There are, however, several disadvantages of basing satellites in high orbits. First, the cost and complexity of satellites and ground stations would increase somewhat, due to tracking requirements (satellites outside GSO orbit more slowly than the earth revolves), and due to the fact that transmitting power and/or receiver sensitivity would have to compensate for the increased distance from the earth. Second, the ability of reconnaissance and surveillance satellites, such as attack warning satellites, to see details on the earth's surface decreases with distance. Third, the cost of launching satellites into orbits above GSO would be somewhat greater, since more energy is required for a given payload mass. Note that the number of satellites necessary to perform a mission need not increase, but that the size, power requirements, and cost of each satellite would be greater.

Passive countermeasures, especially hardening, can go a long way toward decreasing the vulnerability of satellites. They also cause effective ASAT systems to become large, expensive, and detectable. A program of passive countermeasures undertaken now could greatly increase the survivability of future systems, since it would take the Soviet Union at least a decade (and probably much longer) to design, test, and deploy an advanced ASAT capable of threatening critical satellites in GSO. But passive countermeasures are not sufficient by themselves to ensure survivability. There is no known perfect passive countermeasure, nor are there perfect substitutes for space assets. Secure, redundant land systems would not satisfy all strategic requirements and would be very expensive for the U.S. (although less so for the U.S.S.R.). In the case of attack warning, for example, reliance on land-based systems would mean giving up independent confirmation that an attack was on the way, relying on half the current warning time, or giving up reliance on tactical warning altogether, none of which are acceptable.

Active countermeasures

Active unilateral countermeasures are those that threaten an attacking ASAT system. This means deploying one's own ASAT, either to deter attack or to destroy the opposing ASAT. Systems with the latter mission are sometimes called DSATs (defensive satellites), but it is not clear what the technical difference might be between systems that are designed to kill satellites generally and those intended to kill ASAT satellites. A basis for a valid distinction might be DSAT systems that are only effective over a very limited range and that are associated with a specific satellite system. In this case, DSATs could legitimately be seen as strictly defensive, since they could not attack an opposing ASAT system unless it advanced within range. Although this might be a reasonable limitation if imposed in relation to ASAT space mines or missiles, such defenses would be worthless against directed-energy weapons unless the range of the DSAT was at least as great as that of the opposing ASAT, which would make it an ASAT in its own right.

Self-defense systems could not be added to current satellites, since the weight and power requirements of such add-on systems would be far greater than those of the satellite they would be protecting. In addition, the operation of some DSAT weapons could destroy any nearby satellite. A separate DSAT satellite, very much resembling the ASAT systems described above, would be needed. Active defense against a nuclear warhead at a range of 1,000 km, for example, would require a very large NPB or laser. DSAT weapons may not be able to defend the satellite at all against nuclear weapons if they were salvage-fused.

The use of space-based DSATs could create an advantage for preemption and therefore could be crisis unstable. If both sides depend on satellites to perform crucial deterrence functions, and both sides also deploy ASATs to threaten the other's satellites (as well as their ASATs), then substantial benefits could accrue to the side going first. This is essentially the same argument that is used when evaluating vulnerable land-based ICBMs: if both sides have valuable but vulnerable weapons, each will fear preemption by the other, and will therefore be tempted to preempt. A crisis or accident (e.g., collision with space debris) could trigger a satellite war and measurably raise the probability of terrestrial war. Active ASAT defense itself is also likely to be arms-race unstable for similar reasons. If ASATs are practical, then so are DSATs, which could also function as ASATs, leading to an almost inevitable measure/countermeasure arms race in space.

These arguments apply especially to space-based systems; if ASATs are earth-based and do not rely on space-based components, then ASATs could not attack other ASATs, and ASAT deterrence may be crisis stable, though it will still add a component to the arms race. An example is the U.S. ASAT under development, which can reach targets anywhere in LEO but cannot easily be preemptively destroyed.

But if ASATs can be made invulnerable to preemption, then DSATs cannot prevent satellite destruction, they can only threaten retaliation in kind. ASAT deterrence may not work, however, if one side valued the destruction of the other side's satellites more than the survival of its own. This appears to be the case with preemptive strategic attack, where attack warning, communications, and navigation satellites would be much less valuable to the attacker after his missiles were launched, or in the case of a conventional war, in which the U.S. would be more dependent on satellites than the U.S.S.R. Thus, even if ASATs can't threaten other ASATs, the situation could still be unstable.

Bilateral agreements

If both sides have a stronger interest—at least with respect to war prevention—in safeguarding their own space assets (or some part of them) than in maintaining a capacity to destroy the other side's, and if unilateral measures taken to safeguard these assets force ASAT systems to be expensive and detectable, then verifiable bilateral or multilateral agreements to limit ASAT technologies or deployments may be possible and may be perceived by both sides to improve the security of both. For this to occur, policy decisions must be reached on both sides to the effect that such factors as enhanced crisis stability and decreased arms expenditures outweigh the potential wartime advantages to be gained by holding satellite systems at risk. On this reasoning, the Scowcroft Commission—formed by President Reagan to evaluate basing schemes for the MX missile—recommended that the U.S. attempt to negotiate agreements to make critical satellites more survivable.[11]

Measures taken unilaterally to make satellite systems survivable will make bilateral ASAT controls more attractive, because they serve to increase the cost and decrease the effectiveness of ASAT systems, and because they drive ASAT systems to larger dimensions and power requirements so that bans and restrictions are more liklely to be verifiable. A variety of restrictions on ASAT development, testing, and deployment might be considered.

Testing and deployment of the current U.S. and Soviet ASAT weapons, as well as the testing of other earth-based missiles in an ASAT mode, might be mutually restricted or banned. If the perceived military utility of being able to destroy low-orbit surveillance and reconnaissance satellites is great, both sides may want to keep this ability while banning tests at higher altitudes. One might also choose to "grandfather" existing ASATs because of the difficulty of verifying a ban on their deployment (the weapons are small and easily concealable), given that they have already been developed and tested (although the U.S. ASAT is not fully tested).

Verifying restrictions on ASAT testing may be problematic, however. If earth-based ASATs can be tested successfully against a point in space, no one might know that a test had occurred, or that the test involved an ASAT weapon. If current low-altitude ASAT weapons are permitted, might low-altitude tests be sufficient to permit development of a high-altitude ASAT? As an example in which this would be the case, adding a third stage to the current U.S. air-launched ASAT might allow it to reach satellites in GSO; one may want to ban the testing of modifications of current ASATs to prevent this from happening. The importance of these issues should be resolved before designing a specific agreement.

Negotiated "keep-out" zones represent a way to increase the separation between satellites of different nations through formal agreements. An important argument for keep-out zones is they are one of what appears to be only two effective defenses against space mines, whether nuclear or conventional, because of the short range and rapid engagement they imply. Any object that *could* be a space mine could be attacked before it could come within lethal range, an arrangement that is fraught with instabilities. Alternatively, we can try to deal with the problem through the type of agreement proposed here.

As an example of a structured agreement, the United Nations could partition certain zones of space. The most crowded and vital orbit, GSO, is already organized to prevent radio interference. GSO might be divided into 36 zones, each 10° wide, 12 of which would be assigned to NATO and other U.S. allies, 12 to the Warsaw pact and other Soviet allies, and 12 to neutral or non-aligned countries.[12] This type of agreement would have the advantage that only a few existing satellites would need to be moved. Satellites stationed in the middle of a zone would be at least 3,700 km from possible ASATs in an adjacent zone. Even a very large nuclear explosion could not destroy minimally hardened satellites at this distance. Homing missile ASATs parked at this distance would require several minutes for an attack, allowing sufficient time for the threat to be assessed and for a warning message to be relayed to earth.

Other orbits might be organized into, say, 72 spherical shells 5,000 km "thick," starting 10,000 km above the earth and going out to the moon's orbit of 380,000 km, allocated in a manner similar to the 36 zones of GSO. This would include the orbits of NAVSTAR and the Soviet's NAVSTAR-like navigation systems. Space inside 10,000 km and outside the moon's orbit would be unregulated.

Structured keep-out zones may not be acceptable to the world community, however. First, equatorial nations will oppose any measure which deeds the use of the space above their nation to some other country or collection of countries. Second, the current United Nations position is that no state can claim a part of space for its own. Keep-out zones may also impose restrictions on satellite missions. It is current practice for satellites to drift several degrees about their mean positions, which would be impermissible at the edges of keep-out zones. Satellites could be kept on tighter orbits, though this would require more fuel.

There are alternative, less rigidly structured possibilities. Spheres of, for example, 1,000 km radius centered on certain satellites could be agreed upon as keep-out zones. Foreign satellites would be prohibited from entering these spheres without permission. This would appear to allow only 130 or so protected satellites in GSO, but the actual upper limit would be several times larger, since satellites of allied nations could be stationed within each other's zones by permission, and since slightly tilted orbits could be used. Such agreements would work better in non-synchronous orbits, where crowding is less of a problem.

Foreign satellites would have to be allowed the right of friendly passage, subject to certain restrictions, such as a maximum number of transits per day. Satellites from other nations wishing for some reason to share a keep-out zone could be subject to inspection before launch. Counter-rotating GSO satellites and intersecting elliptical orbits pose a problem and might have to be limited or banned.

The fact that many Soviet attack warning and communications satellites are in Molniya orbits rather than GSO may be very important when designing a keep-out zone agreement. These orbits are inclined and highly elliptical, with an apogee (highest altitude) of approximately 40,000 km and a perigee (closest approach) of 500 km. Satellites in this orbit spend over 90 percent of their time on one side of the earth, where they function like satellites in high circular orbits. But the fact that they come so close to the earth presents a special problem for keep-out zone agreements of all kinds, especially because of the asymmetry between the U.S. and the Soviet Union in the number of satellites in this orbit. The value of keep-out zones is greatly diminished for satellites in Molniya orbits, because earth-based ASATs would be much more effective against them. Air-launched ASATs, such as the current U.S. ASAT, can be launched from the Southern hemisphere and therefore can attack these satellites at perigee, though to destroy an entire satellite system one would have to wait for all satellites to pass through perigee, which would take up to 12 hours.

Bans on testing large (e.g., greater than one megawatt) ground-based lasers in an ASAT mode and on the testing and deployment of large mirrors in space can mitigate the prompt threat from earth-based systems. Many of the remaining ASAT threats—space-based kinetic-energy and directed-energy weapons—could be ameliorated by a ban on their development, testing, and deployment in space. The testing and deployment of nuclear weapons (and X-ray lasers) in space is already banned. With the exception of nuclear-weapon-driven devices, such a ban should be verifiable since the weapons in question would be large and identifiable, and since they would require testing in space to be reliable, which could be detected.

Finally, if nuclear directed-energy weapons, such as the X-ray laser, are particularly worrisome, a comprehensive test ban treaty or a low-yield threshold test ban treaty could be negotiated to inhibit their development and testing.

Verification

A variety of technologies can be used to aid in verifying the types of agreements proposed here. Improved space tracking and surveillance systems are usually proposed in connection with ASAT arms control, since they would aid greatly in verifying compliance with keep-out zones, restrictions or bans on ASAT testing, and bans on ASAT deployment in space. These systems can cut both ways, however, in the sense that a sophisticated space surveillance system can form the basis for an ASAT system as well as safeguard existing satellites, much as large terrestrial radars can form the basis for an antiballistic missile (ABM) system as well as an early-warning system. The current U.S.-U.S.S.R. ABM treaty prohibits large radars that are not on the perimeter of the nation and looking outward to prevent quick treaty break-out. Just as in the ABM case, the construction of large space surveillance systems may cause concern about compliance with an ASAT treaty.

It may be possible to limit space tracking systems to the mission of verifying compliance with agreements, so that such systems would not be capable of missions that threaten these agreements. For instance, keep-out zones could be monitored by infrared sensors on board the protected satellite, rather than by extensive networks of dedicated satellites which could easily form the basis for an ASAT weapon system. This is a area that should receive more thought.

Space tracking systems can be supplemented by heat, X-ray, and acceleration sensors on board U.S. satellites that verify that the satellite is not being attacked by lasers, NPBs, nuclear weapons, or missiles. Such sensors are a valuable countermeasure to ASAT warfare with or without arms control.

The deployment of nuclear weapons in space (including nuclear space mines, nuclear homing missiles, and X-ray lasers) is banned by the Outer Space Treaty. At present, it is not possible to verify compliance with such a ban. In theory, one could inspect satellites before launch or while in space to detect radiation from the fissionable materials, but relations between the U.S. and the U.S.S.R. have not been conducive to such arrangements in the past (although there recently have been hints that the U.S.S.R. may be willing to change its position). The threat from nuclear space mines might be sufficiently defused by keep-out zones, provided that the space surveillance system used for verification can determine if keep-out zones have been violated.

Nuclear explosions in space are banned by the Limited Test Ban Treaty. This ban is verified by nuclear-burst detection systems on board current U.S. satellites, for distances at least out to the moon's orbit. This verification capability could be extended to much deeper orbits by deploying satellites dedicated to this function.

Verifying a ban on the testing of ground-based lasers in an ASAT mode is, in principle, fairly straightforward. As noted above, lasers powerful enough to pose a threat to high-altitude satellites will be very large. The U.S. has shown the ability to find much smaller lasers. Ground-based lasers are at fixed locations, and can only be tested during cloudless periods, when space surveillance of the lasers is also possible. By posting surveillance satellites over the laser sites, one should be able to detect, by the scattering of light as the beam passes through the atmosphere, whether the laser is being tested in an ASAT mode. In addition, if a comprehensive space surveillance system is available, the optical signals of all large space objects could be monitored, which would make it possible to determine if they were being illuminated by a laser. If necessary, on-site sensors could be placed at large laser installations to ensure treaty compliance.

It is possible to verify a ban on space-based lasers and particle beam weapons, since such weapons would have to be large and distinctive if sized to attack satellites several thousand kilometers away. Testing of such weapons would be easy to detect by observing their thermal signature, or by detecting effluents given off during their operation. The destruction of target satellites by kinetic-energy weapons can also be verified by space tracking and surveillance systems.

Finally, it should be noted that the U.S. must maintain an excellent space intelligence system with or without arms control. In general, it is much easier to detect and monitor certain activities when they are constrained rather than widespread. The surveillance technologies necessary to verify the sorts of agreements outlined here are much less ambitious than those envisioned for the Strategic Defense Initiative (SDI), for example.

IMPACT OF ASAT LIMITATIONS

What impact would bilateral agreements such as those described here (coupled with unilateral actions to make the satellites as surviva-

ble as economically practical) have on the military policies of the
United States and the Soviet Union? The most obvious effect would
be to deny to both countries the ability to destroy high-altitude
satellites, both those that are essential for deterrence (attack warning
and communications), and those that may not be (navigation and
nuclear burst detection). The development of new technologies, such
as earth-based HELs, would be allowed, but they could not be tested
in an ASAT mode. Space-basing of such technologies would be banned
altogether.

SDI

The agreements considered in the previous section would impose
restrictions of relevance to many military missions that are not now
performed in space, but could be, such as ballistic missile defense.
Any ABM system comprised of directed-energy weapons that are
powerful enough to destroy thousands of missiles during a few
minutes (boost phase) or tens of thousands of reentry vehicles (RVs)
over tens of minutes (mid-course phase) would almost certainly be a
threat in principle to satellites in GSO. Such systems could be more
effective as ASATs than as ABM weapons, since the ASAT mission can
be performed at the moment of one's choosing, and since satellites
are in general softer targets which travel on predictable paths.

Critics of this view point out that the distances involved in attacking
a satellite in GSO are much greater than in attacking a missile:
roughly 36,000 km versus 1,000 km, the commonly-assumed orbit
for the space-based laser battle stations that might form the core of
an ABM system in space. Since the intensity of a laser decreases as
the square of the distance, the intensity at GSO will be over 1,000
times less than that achieved on a targeted booster; since satellites can
be made almost as hard as boosters, satellites in GSO should be safe
even from a system potent against missiles. The SDI Organization
supports high-orbit ASAT arms control measures, based on the premise that many of the sensors necessary for strategic defense would be
placed in high orbits.[13]

This analysis is flawed in several ways, however. First, lasers could
irradiate satellites for a much longer amount of time than they could
irradiate boosters, because there are many fewer satellites than boosters. ABM systems will have to shoot successfully at 1,000 (boost-
phase) to 10,000 (mid-course) objects per minute, while ASAT systems
would only need to attack at most 100 satellites in a few minutes. In
addition, many more laser battle stations could participate in an ASAT
attack than in an ABM defense, because each laser satellite could
target at least half of all high-orbit satellites, while in the ABM case,

less than 1/10th of the laser satellites can target missiles launched from a point on the earth. These two considerations nullify the effect of the added distance, and make a space-based laser system at least as effective as a high-altitude ASAT as an ABM system. In fact, an ABM system would probably be more potent as an ASAT since satellites are more difficult to harden than boosters or RVs, and since an attack can usually be coordinated better than a defense and can occur at a time of the attacker's choosing. Earth-based laser ABM weapons should be particularly effective against satellites, since in most such schemes the laser energy is reflected from mirrors in GSO to boosters at low altitudes. If the laser beam could destroy boosters after going up to GSO and back again, then it could obviously be even more effective against objects in GSO.

Directed-energy systems capable of boost-phase or mid-course ballistic-missile defense would threaten the high-altitude space assets that the agreements discussed above seek to protect. The development of countermeasures may or may not change this situation. Moving to deep space orbits, for instance, could be effective. But a defense based on technologies with limited range (small homing missiles launched from space platforms, for example) might not have much effect on satellite systems. The interactions between defenses and space system security will have to be considered very carefully.

How can satellites be protected in a world of space-based ballistic missile defenses? Obviously, hardening and active defense (shooting back) will become even more important. What role bilateral agreements can play depends a lot on whether or not the transition to such defenses is cooperative, as many members of the Reagan administration insist it must be. If defenses are deployed cooperatively, meaning not only that defenses are viewed as desirable and stabilizing by both sides, but even that technologies or systems may be shared, then there is no reason why both sides cannot mutually agree to limit offensive countermeasures to such a defense, including ASAT weapons. Keep-out zones could become, by agreement, self-defense zones. SDI technologies could be designed so they would not threaten the opponent's system with preemptive destruction, and critical satellites could be moved to orbits high enough to avoid the threat posed by these missile defense weapons.

If, on the other hand, the transition to defense were not cooperative (which is much more likely), then no controls on ASAT weapons seem possible, for the simple reason that ASAT weapons would be a principle countermeasure to space-based ABM system components. For every ABM system deployed, the opponent would want to deploy an ASAT capable of negating it.

Although one cannot protect present and currently planned satellite systems through restrictions on ASATs and deploy an ABM system at the same time, this does not mean that ASAT arms control is impossible while research continues on SDI. Resolution of this question depends on whether demonstrations and field experiments that may of themselves provide some ASAT capabilities are deemed necessary to the SDI R&D program. If, for some significant period of time, these demonstrations and field tests are not deemed necessary, ASAT limitations could in theory at least be agreed on for that period. In the final analysis, however, developing SDI and ASAT arms control are probably incompatible.

Space as sanctuary

Limiting the threat to high-altitude satellites would create a sanctuary in space to a certain extent. It is possible that this would actually encourage the use of space for military, but non-weapons, systems, since they would be safe from attack. Although it is difficult to foresee examples of this type of behavior, one can imagine that, for instance, a device could be built that could track ballistic-missile submarines from space. Another example might be high-altitude surveillance satellites that could tell if targets were destroyed by an initial nuclear attack, in order to direct additional missiles to the surviving targets. These developments would represent a clear threat to deterrence, and we would certainly want the capability to destroy such satellites. But it seems premature to forego arms control just because of these theoretical possibilities.

Space debris

A side benefit of a ban on ASAT testing would be the reduction of space debris. The debris accumulating in orbit in the absence of ASAT testing—paint chips, pieces of exploded boosters, etc.—is suspected to have damaged several satellites in the past few years. The risk of collision of a shuttle-sized object with a large (greater than one centimeter in diameter) piece of debris is currently about 1 percent per year.[14] It has been estimated that the planned U.S. ASAT tests could double the total amount of debris in LEO.[15] A more extensive ASAT testing program against real satellites might make whole regions of space unnavigable. The problem would become critical if ASAT tests were conducted at high altitudes, since debris there is removed much more slowly—indeed, at geosynchronous altitude, orbits decay only one kilometer in altitude every one thousand years.[16] ASATs could be tested against balloons, which might decrease somewhat the amount of debris generated, but not completely, since such balloons are likely to be heavily instrumented.

CONCLUSIONS

The attack warning and communications satellites based in geosynchronous orbit are essential for the maintenance of deterrence. These satellites are intended to perform vital functions during, or at the outset of, a nuclear war. Although current earth-based weapons do not pose the threat of rapid destruction of high-altitude satellites, a variety of future technologies could be so employed as anti-satellite weapons. These include powerful earth-based lasers, and space-based mines, kinetic-energy weapons (rail guns and homing missiles), and directed-energy weapons (particle beams, optical lasers, x-ray lasers, and microwave weapons).

A variety of passive unilateral countermeasures, including maneuvering, use of decoys, hiding, redundancy, and especially hardening and the use of deeper orbits, can go a long way toward making satellites more survivable and forcing ASATs to become more sophisticated and costly. A vigorous program of passive countermeasures should be undertaken, and it should not be delayed by progress in space arms control since such measures will make agreements more robust. But passive countermeasures cannot by themselves assure survivability. None can prevent the destruction of satellites from, for example, an X-ray laser or a space mine placed near the satellite.

Active defense of satellites may be able to protect satellites and thwart the emplacement of space mines, but this is likely not to be crisis stable. Each side could fear preemption and require a hair-trigger posture to prevent both satellites and ASATs from being destroyed. Active defense would probably also lead to a measure/countermeasure arms race as each side attempted to make its satellites and ASATs invulnerable to the other's ASATs. In any case, ASAT deterrence through retaliation in kind is not likely to work in a number of war-time situations, because the destruction of an opponent's satellite systems is more valuable to an attacker than preservation of systems.

Bilateral or multilateral agreements can play a large role in safeguarding high-altitude satellites. Bans on testing (and in some cases deployment) of new technologies in an ASAT mode can complement negotiated keep-out zones in limiting the threat from earth-based as well as space-based ASATs. Large-scale testing of an advanced ASAT would be observable, so a ban would be verifiable.

Limitations on ASATs are not likely to be compatible with the deployment of a strategic defense with space-based battle stations. Although such battle stations would have to operate over a much larger distance when used in an ASAT mode, the scale of an ASAT attack is so much smaller than that of a missile defense, and so many more battle stations can participate in the ASAT attack than in the missile defense, that a successful ABM system (other than terminal or other limited range defenses) would be a potent ASAT. It is unlikely that either country would willingly forego the ASAT option if the other had plans to deploy space-based ABM components, since ASAT weapons may be one of the most effective countermeasures against such a system.

These assessments lead to the inescapable conclusion that the United States now faces critical strategic decisions about the future of our military satellite systems. The outcome of these decisions is likely to affect not only the cost, effectiveness, and reliability of those systems, but also the entire framework of our national security programs and policies.

NOTES

1. For a review of military space systems, see C. Richard Whelan, *Guide to Military Space Programs*, Pasha Publications, 1984.

2. All 24-hour (geosynchronous) circular orbits are about 36,000 km above the earth's surface, and those that are not inclined with respect to the earth's equator (GSO) hover above the same point on the equator. It is this unique property that causes so many satellites to be located in GSO. For a review of satellite orbits, see Ashton Carter, "Satellites and Anti-Satellites: The Limits of the Possible," *International Security*, Spring 1986.

3. Space Analysis and Data Division, "Space Computational Center Satellite Catalog," North American Air Defense, April 1982.

4. R. L. Garwin, K. Gottfried, and D. L. Hafner, "Antisatellite Weapons," *Scientific American*, June 1984.

5. James B. Schultz, "Space System Designs Promote Survival of the Fittest," *Defense Electronics*, June 1985.

6. U.S. Congress, Office of Technology Assessment, *Anti-Satellite Weapons, Countermeasures, and Arms Control*, OTA-ISC-281, U.S. Government Printing Office, September 1985.

7. *Ibid.*

8. R. Jeffrey Smith, "Space Experts Challenge ASAT Decision," *Science*, May 18, 1984.

9. One kiloton is the amount of energy released by 1,000 tons of high explosive. A 15 kt weapon destroyed Hiroshima, and most weapons in the U.S. stockpile have a yield of a few hundred kilotons.

10. Donald J. Kessler and Shin-Y Su, Eds., *Orbital Debris*, NASA Conference Publication 2360, March 1985.

11. *Report of the President's Commission on Strategic Forces*, April 6, 1983.

12. Albert Wohlstetter and Brian Chow, "Arms Control that Could Work," *Wall Street Journal,* July 17, 1985.

13. R. Jeffrey Smith, "Limited ASAT Proposal Gains Backers," *Science*, May 18, 1984.

14. Kessler and Su, *op. cit.*

15. Eliot Marshall, "Space Junk Grows with Weapons Tests," *Science*, October 25, 1985.

16. Kessler and Su, *op. cit.*

AMERICA'S RETURN TO SPACE: U.S. SPACE TRANSPORTATION POLICY

Edwin A. Deagle, Jr.

LOOKING AHEAD IN early 1985, an informed observer would have been hard pressed to imagine a set of circumstances that would deny American access to space completely for two full years. Yet in less than twelve months such circumstances unfolded, with literally catastrophic results: critical national security space launch needs would not be met for two to three years; some of NASA's interplanetary scientific missions faced postponement of five years or more; and foreign and commercial satellite operators began looking overseas in search of launch capacity. All of this took place just as the international competition for scientific and commercial leadership in space for the rest of the century was beginning in earnest. In the short history of American activities in space it is difficult to find a more monumental failure in U.S. policy.

The specific events that brought America to this dismal state of affairs over the course of a year are now painfully evident. First was the failure of the Air Force's Titan 34D—7th of the 15 funded—in August 1985. The second incident was the loss of the shuttle Challenger, tragically important both because of the loss of life and because of its exposure of serious flaws in U.S. space policy. The third incident occurred with the destruction of the 8th Titan 34D in April 1986 a few seconds after lift-off, effectively grounding the only U.S. expendable launch vehicle (ELV) supplier with an operating assembly line. In May 1986 the loss of a Delta carrying the NOAA GOES weather satellite grounded that system as well. Finally, the failure of an Atlas Centaur following lightning strikes in March 1987 grounded the only remaining U.S. launch system.*

---

*Disaster was not confined to the U.S.; in May the 18th flight of France's Ariane failed, grounding that system until September 1987. Early in 1987 the U.S.S.R. lost two Proton launches due to fourth stage failures.

27

After May 1986, the U.S. was again successful, with its remaining ELV stockpile: in September Delta flight number 179 was launched flawlessly, carrying an SDI experimental payload, and an Atlas E successfully carried the replacement GOES to its assigned orbit. In December an Atlas Centaur launch placed a Navy VHF communications satellite in orbit (but the subsequent Atlas Centaur failed and the remaining launch vehicle was badly damaged on the ground in 1987). In 1987 NASA successfully launched two Deltas and the Air Force launched a Titan 34D.

Following the report of the Presidential Commission on the Space Shuttle Accident, the U.S. government began a shift in policies designed to recover U.S. access to space. The principal features of the new policy initiatives are:

• A decision to procure a replacement orbiter.[1]

• A U.S. Air Force space launch recovery plan to expand its procurement of Titan IV from 10 to 23 and to procure 20 Delta medium launch vehicles (MLV) to launch NAVSTAR satellites. The Titan IV and Delta II are seen by the Air Force as complementary to the shuttle and providing in conjunction with it assured access to space.

• A 1987 Air Force plan to procure an advanced launch system (ALS), a heavy launch vehicle to support SDI and the space station.

• An effort to encourage development of a U.S. commercial ELV industry by requiring that the MLV be capable of providing commercial satellite launch services, and simultaneously, barring most commercial satellites from shuttle launch.[2]

Given the magnitude of the current crisis in U.S. access to space, it must be asked whether the policy initiatives outlined above constitute an effective and sufficient response. The answer to that question depends on an assessment of four points:

• U.S. space transportation policy before the Challenger accident and the impact of that policy on the current space launch crisis;

• Demand for launch capability during the critical period 1988 to 1995;

• The main determinants of U.S. launch vehicle supply; and

• The appropriate mix of U.S. policies regarding space transportation programs, given the supply and demand for launch services and current space policy objectives.

U.S. SPACE TRANSPORTATION POLICY PRIOR TO THE CHALLENGER ACCIDENT

The centerpiece of U.S. space transportation policy following the successful Apollo program has been the development and operational use of the space transportation system (STS)—the space shuttle. Since 1972 the United States has invested about $40 billion in the shuttle program and it is the largest component of NASA's budget (about 25–30%).

A distinguishing feature of the shuttle program had been its ambitious objective of moving quickly beyond the research and development phase to full scale, routine operations as a "space trucking company." In its original STS Baseline Operations Plan of November 20, 1978 NASA defined operational requirements for the shuttle system: 6 flights the first year, 15 the second, 24 the third and by the sixth year 60 flights.[3] Aside from the technological optimism implied, the proposed flight rate reflected the need to recoup the very high fixed-cost investment made to develop the shuttle system—estimated by the Congressional Budget Office to be $25 billion.[4] Even today the full cost of each shuttle flight is extremely sensitive to the flight rate, as Table 1 indicates.

NASA is not required to recover full costs for launch services provided by the shuttle, but is expected to recover the operational out-of-pocket expenses associated with launch services to non-NASA customers. NASA's pricing structure for shuttle launches reflects both an appreciation of launch services revenue as a contribution to its budget and the need to compete with other launch services suppliers, especially U.S. ELV suppliers and France's Arianespace. In general shuttle launch prices have been substantially below U.S. ELV launch costs. Arianespace until 1986 set its launch prices to be roughly equal to shuttle prices.

In the 1970s as the shuttle system was being developed, NASA sought to achieve—implicitly, at least—the status of a monopoly supplier of launch services. Initially it was very successful. After a

TABLE 1

FULL COST PER SHUTTLE FLIGHT, 1989
(FY 86 $)

	Number of Flights/Year		
	24	18	12
Full Cost/Flight	$208M	$258M	$358M

Source: Congressional Budget Office

long and bitter battle the Defense Department was instructed to shut down its procurement of ELVs and purchase launch services exclusively from NASA's shuttle.[5] At the same time NASA proposed launch prices ($74 million for the full bay)* that undercut the prices U.S. ELV manufacturers would have to charge in order to compete for launch of foreign and commercial communications satellites.

In effect NASA became internationally competitive in providing space transportation by developing a large government captive market (Defense and NASA representing about 75 percent of the non-Communist launch market), and by seeking to become economically efficient as the shuttle flight rate rose.

As a market strategy, the NASA plan was a brilliant success. Despite competition from Arianespace, levelling of demand for launches from the commercial satellite market, and debilitating cost growth and schedule delay in the shuttle program, NASA by 1985 had secured a virtual monopoly position for the shuttle system in the traditionally U.S.-dominated launch services market. Of the relatively firm foreign and commercial satellite launches booked in November 1985, NASA had contracted for forty and Arianespace eight.[6] Moreover, the shuttle had a base of 4 to 8 flights per year scheduled for the Defense Department from 1986 to 1992 and an equivalent—but much less firm—set of NASA bookings.[7]

From technology management and budget points of view, the NASA shuttle strategy was clearly a disaster, as recent events vividly demonstrated. NASA's monopoly supplier strategy early and continuously conflicted with its Apollo-derived traditions of technical conservatism and with contemporary budget realities. Ironically, budget austerity assured that market strategy dominated NASA decision-making until technical and budget difficulties—and thus the Challenger accident—exposed the underlying reality.

A useful way to examine that reality is to ask what might have happened had the Challenger accident not taken place when it did, but two years later, in 1988. The Titan and Delta (and Ariane) ELV failures would have accelerated the U.S. move toward a shuttle-based launch strategy. By 1988 it is doubtful that any U.S. ELV manufacturers (other than Martin Marietta) would have remained in the business in the face of NASA and Arianespace subsidized competition. The remainder of the ELV technical and human resource base by then would have been dismantled. One can only speculate about what would have happened to the shuttle flight rate, given what has emerged in the aftermath of the Challenger accident.

*In FY 1982 Dollars

The result of a shuttle accident in 1988 would have been much more severe than in 1986. The United States would have faced not only a twenty-four month stand down of the shuttle, but also the prospect that restoration of a mixed shuttle-ELV fleet would take three to five years—just as the need for massive launch capability in support of the space station (not to mention SDI) emerged. The present communications satellite "transponder glut" likely would have evaporated by 1988 and satellite operators by then would be facing an urgent need to replace their space assets.

The critical point is that the Challenger accident—painful to the country though it was—was not the source of the United States' present difficulties in space. The culprit is a space transportation policy that was fated to produce catastrophe because it could not absorb the risk of a single failure. Ironically, if there is a preeminent principle usually applied in the design of space technology, it is the avoidance of systems vulnerable to single point failures.

FUTURE DEMAND FOR LAUNCH SERVICES

In general estimates of the demand for launch services have been high. In 1977 NASA estimated that 572 shuttle flights would be flown between 1980 and 1991. By 1980 that estimate was reduced to 487 and by 1985, to 165 flights, or 30 percent of the 1977 estimate.[8] NASA's forecasts of overall non-Communist launch activity for 1988, for example, continually have been scaled down (See Figure 1).

There are several reasons for the over-estimation of launch demand. First, in response to launch projection surveys, potential new entrants in the commercial satellite industry regularly underestimate the technical, market and financial constraints that together form a formidable barrier to entry in the deregulated telecommunications marketplace. It is one thing to commission market surveys, apply for orbital slots and deposit $100,000 to reserve a shuttle or Ariane launch. It is quite another to secure $150–500 million for serious entry in a hotly competitive satellite communications market that is in the midst of a severe shakeout from overcapacity and competition from fiber optic cable systems. Of the 69 foreign and commercial payloads booked on Ariane and the shuttle before the Challenger accident, for example, perhaps less than thirty will actually fly.[9]

A second reason for the over-estimation of demand is that communications satellite customers—especially the Defense Department—have been assuming that satellite life spans are shorter than actual experience has demonstrated. Further, in the case of commer-

FIGURE 1
NASA FORECASTS OF LAUNCH ACTIVITY IN 1988

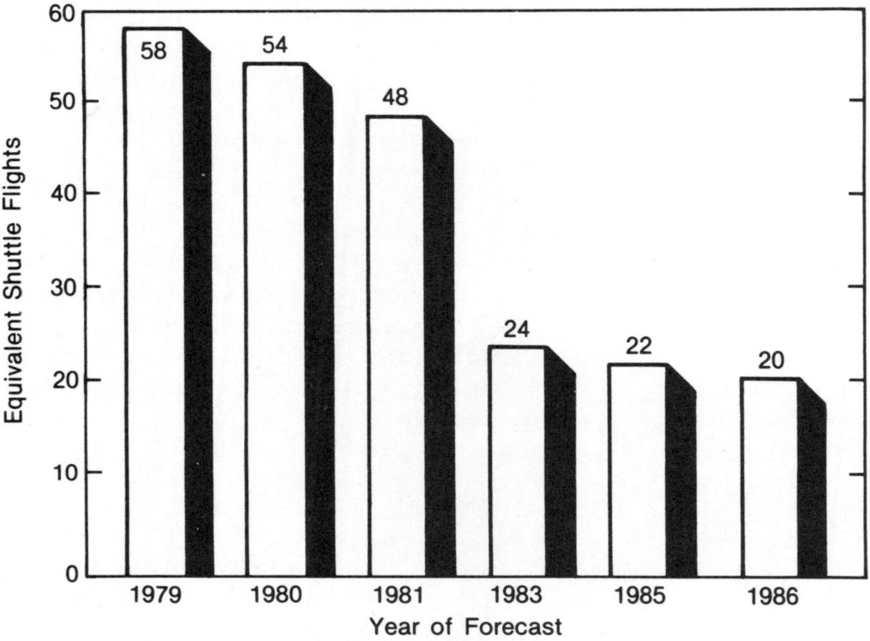

SOURCE: Congressional Budget Office, based on NASA data.

cial satellites, the larger and more capable new class of satellites can replace older systems on a two-for-three or one-for-two basis. Both factors tend to lower launch demand. Finally, NASA's projection of its own launch demands are usually overstated because Congress consistently has provided lower appropriations than envisioned in NASA planning. Given the cost of reconstituting the shuttle system it is doubtful that this pattern will change appreciably in the future.

Launch demand during the period 1988–1990 will be comprised primarily of the critical parts of the backlog that has emerged from 1986 to 1988. Estimates of the size of that backlog, like estimates of future launch demand, depend heavily on assumptions. In October 1986 the Congressional Budget Office (CBO) prepared an analysis of launch demand and supply. In addition to the official NASA launch demand projection CBO presented two alternatives: A *constrained case*, in which the official projection was downscaled to correct for the budgetary and technical optimism that usually exaggerates federal requirements; and an *historical case*, a linear projection of the previous sixteen years of experience from 1970 to 1985.[10]

For its constrained case CBO reduced Defense demand to 70 percent of the official projection (in the past NASA has actually flown about 70 percent of projected Defense payloads, year by year). NASA and other government agency launch demand was reduced by 50 percent—a rate slightly above that experienced in the past 15 years. Commercial demand was reduced by 50 percent to reflect emerging international competition for launch services. The historical case is a simple extrapolation of the actual trend of the past sixteen years, modified to exclude the high launch activity of the Apollo program in the early 1970s. The official projections and CBO's alternative projection are summarized in Table 2.

CBO's estimates for the foreign and commercial satellite launch demand available to U.S. ELVs and the shuttle may be high. The author's estimates project an average of nine communication satellite launches per year for the *entire* non-Communist foreign and commercial market, roughly equivalent to three shuttle launches. Only half or less of that demand might fall to the shuttle and U.S. ELVs.

In hindsight launch demand projections are often seen to be wrong, and it would appear that the official case is very optimistic. In conjunction with projections about future launch capacity (addressed next) lower-than-expected launch demand has important ramifications for U.S. space transportation policy.

FUTURE SPACE TRANSPORTATION CAPACITY AND SUPPLY

Much more than is the case with launch demand, the availability of launch capacity is a function of U.S. government policies. There are two reasons why this is true: First, government programs constitute the largest component of launch demand, and the nature of that demand traditionally has dictated the requirements for launch capac-

TABLE 2

AVERAGE ANNUAL LAUNCH DEMAND 1990–2000

(equivalent shuttle flights)

Case	DoD	NASA & Other U. S. Gov't.	Commercial	Total	1986–88 Backlog
NASA	13.1	10.9	6.9	30.9	60.0
CBO					
Constrained	9.2	5.5	3.3	17.9	32.5
Historical	N/A	N/A	N/A	10.9	28.0

Source: Congressional Budget Office

ity. As a result, commercial satellite designers have designed space-craft to match the launch capabilities provided by industry to the government.

The second reason for the importance of government policy to launch capacity is that the fixed costs of developing complex rocket launchers and the supporting physical infrastructure are so enormous relative to market return that private capital has never been committed without the assurance of firm government contracts. Thus the emergence of a U.S. aerospace industry devoted to the production of ELVs (and later the shuttle) developed around government procurement of vehicles and supporting services. In effect the U.S. government acted as a monopsony consumer of launch services and dominated launch service supply—both for itself and for foreign and commercial customers. No private U.S. company challenged the government's position for a stake in the smaller commercial satellite market.

But events have taken place in recent years which have undermined U.S. government/private industry arrangements for the provision of launch services. The first was the creation of Arianespace, a quasi-public enterprise one-third owned by the French National Space Agency and two-thirds owned by European banks and aerospace firms, and chartered to provide launch services with its Ariane ELV series on a commercial basis. Even before the Challenger accident Arianespace was well on its way to becoming a stunning commercial success; in January 1986 it had a backlog of 28 satellite launch contracts worth $1 billion.[11]

The second event was passage of the Commercial Space Act of 1984 and a prior Presidential directive (NSDD 94) supporting the commercialization of the U.S. ELV industry. In part the new policy resulted from the free enterprise philosophy of the Reagan administration. But it also reflected the concerns of the U.S. ELV industry, about to be put out of business by sole U.S. reliance on the shuttle; the example of Arianespace which proved that an international commercial market existed; and the concerns of satellite operators who increasingly found the launch delays and manifest uncertainties plaguing the shuttle damaging to business plans.

The U.S. Defense Department, which had been waging a battle for a mixed shuttle-ELV fleet, took the President's directive seriously and implemented changes to procurement regulations to allow U.S. ELV companies use of government launch facilities at marginal cost. NASA took token steps to sign provisional launch service purchase agreements with U.S. ELV companies but, given its overriding concern about use of the shuttle, made little real effort to support commercialization. Before the Challenger accident no U.S. ELV company had a firm contract to provide commerical launch services.

The third event was the international response to the grounding of the shuttle fleet and the Titan and Delta in early 1986. As the magnitude of the disaster became clear so too did the folly of reliance on a single U.S. means of access to space. The subsequent grounding of Ariane and the near-term saturation of the Arianespace manifest prompted other countries with ELV capabilities to approach the international market for launch services. The chief beneficiary thus far is the Chinese Long March booster which has launch reservations with Western Union and PanAmPacificSat and is looking for others. The USSR, too, seeks launch contracts, for its Proton launches, and Japan can be expected to enter the market with its H2 launchers early in the 1990s.

In response to the launch crisis the Reagan administration has taken two key steps thus far. The first was to seek a more balanced fleet of launch capability, through procurement of a replacement orbiter and as noted, an Air Force program of procurement of thirteen more TITAN IV ELVs (for a total of 23), refurbishment of TITAN II ICBMs to launch small payloads, and procurement of 20 Delta II medium launch vehicles (MLV) for its NAVSTAR navigation satellites. In further pursuit of NSDD 94, the Air Force has required that the MLV be capable of launching commercial satellites in competition with Ariane. The administration's other new policy, designed to reinforce its ELV commercialization initiative, was to bar most of the commercial satellites with pre-Challenger shuttle launch contracts from the new 1988–90 shuttle manifest.

The intent of these new policy initiatives seems quite clear. Barring commercial satellites from the shuttle reduces scheduling pressure on the orbiter fleet and provides a new market for the U.S. ELV industry struggling to enter the international commercial launch services market. Air Force procurement of additional Titan IV ELVs and the new Delta II MLV provides a base of Defense Department launch vehicle demand to further aid U.S. ELV commercialization and, at the same time, provides a more balanced U.S. launcher fleet.

Whether these policies are sufficient depends on a detailed comparison of launch supply with launch demand. At first glance it would appear that U.S. launch capacity—through both the shuttle and ELVs—will be very constrained relative to demand well into the 1990s. A recent assessment of shuttle flight rates by a National Research Council (NRC) panel casts substantial doubt about the wisdom (and feasibility) of NASA's planned shuttle flight rate.[12] Determinants of the shuttle flight rate include the number of orbiters, availability of spare parts, capability of the logistic support system, and personnel and facilities for turnaround and training. These factors affect the average turnaround time and thus dictate the overall flight rate of the shuttle fleet.

NASA has announced a planned annual flight rate of 16 for the four-orbiter fleet, which translates into four flights per orbiter per year and a turnaround time of sixty-five workdays* The average shuttle processing time at Kennedy Space Center is estimated by the NRC panel to have been seventy-five days. The panel concluded that a sustainable turnaround time in the future would lie in the sixty to seventy-five day range.[13] Those figures yield a flight rate per orbiter of 2.7 to 3.3 and a four-orbiter fleet annual flight rate of ten to twelve. In its report CBO estimates a four-orbiter fleet rate of twelve to sixteen.[14] Thus, shuttle capacity as a significant component of overall U.S. launch capacity may well be less than NASA forecasts.

However, when compared to overall demand and given the growing availability of ELVs, the mismatch between launch supply and demand may disappear. CBO's analysis makes this point very clearly. Table 3 displays CBO's estimate of U. S. space launch capacity in 1989 and the effect of options for expanding that capacity.

Arrayed against the CBO estimate of launch demand (the constrained and historical cases), available capacity would be more than adequate without procurement of a fourth orbiter or additional ELVs (see Table 4).

Thus, from the perspective of launch demand and supply capacity, it would appear that current U.S. space transportation policy as articulated by the Reagan administration is adequate and appropriate, and vastly superior to pre-Challenger policies. It is not clear, however that the specific policy steps undertaken to implement the transition from the old space transportation regime to the new will produce the intended results. The analysis that follows will attempt to show that those steps may not produce the intended results and in fact may produce the opposite. To be specific, the United States in the early 1990s might have neither assured access to space nor a viable domestic ELV industry.

FUTURE U.S. SPACE TRANSPORTATION POLICY

The starting point for U.S. space transportation policy has to be a precise specification of the role of the shuttle. The shuttle has a

---

*Assuming a three-shift, five-day work week. In 1977 NASA estimated a turnaround time of fourteen days.

TABLE 3
U. S. SPACE LAUNCH CAPACITY IN 1989 AND EXPANSION OPTIONS
(in equivalent shuttle flights)

Launch Systems	Equivalent Shuttle Flights per Year
1989 Capacity	
Orbiters	9–12
Expendable Launch Vehicles	
Titan Family	9
Medium Launch Vehicle	3
Subtotal, ELVs	12
Total	21–24
Expansion Options	
New Orbiter	3–4
Additional ELVs	3–5

Source: Congressional Budget Office.

TABLE 4
ALTERNATIVE DEMAND PROJECTIONS COMPARED WITH ESTIMATES OF CURRENT POLICY AND EXPANDED CAPACITY
(In equivalent shuttle flights)

	Projected Annual Average Demand	Estimated Current Policy Capacity	Estimated Expanded Capacity (Replacement orbiter)	Projected Demand as a Percentage of Current Policy Capacity	Projected Demand as a Percentage of Expanded Capacity
Official Case	28.5	21–24	24–28	119–136%	107–119%
Constrained Case	16.5	21–24	24–28	69–79%	59–69%
Historical Case	10.5	21–24	24–28	44–50%	38–44%

Source: Congressional Budget Office

unique and extraordinarily valuable role in relation to manned space missions. Given the likely pace of development of the national aerospace plane, the shuttle will carry the burden of manned space missions well into the next century. This fact has important consequences for the size of the shuttle fleet, its flight rate and what payloads it should carry.

At the heart of this issue is the reliability of the shuttle system. The demonstrated reliability thus far is twenty-four successes for twenty-five flight attempts, for a reliability of 96 percent. As R. P. Feynman pointed out in the Rogers Commission report, system reliability will probably prove to be higher than that, but it will not be 100 percent; his estimate is in the range of 99 percent.[15]

As Table 5 indicates, the probability that there will be no more shuttle accidents through the life of the system, given system reliability of 96–99 percent, is very small, and is unlikely to permit design aspirations of 100 flights per orbiter.

If the United States plans a shuttle capability through the turn of the century (entailing perhaps 100 flights or more) it must procure replacement orbiters at frequent intervals. Further, the flight rate must be designed to assure that true system reliability is closer to 99 percent than 96 percent or the orbiter inventory can be expected to erode rapidly. Moreover, because the shuttle must be man-rated, an accident will invariably result in a lengthy stand down, as the current situation vividly demonstrates.

Thus the shuttle system must be husbanded as a very scarce, unique and fragile national resource. It cannot be a space truck. Nor can it be the primary United States means of access to space. Its use must be confined to manned space missions, flown at a rate low enough to assure very high system reliability.

This conclusion has important consequences for the aspect of U.S.

TABLE 5
PROBABILITY OF NO FURTHER SHUTTLE ACCIDENTS

	True System Reliability	
	96%	99%
10 Flights	66%	90%
20 Flights	44%	82%
100 Flights	1.7%	37%

$$\text{NOTE: Probability} = \left(\frac{\text{System Reliability }\%}{100}\right)^n$$

n = Number of flights

space transportation policy concerned with assured access to space. The present policy—which envisions the shuttle system as the primary means of access to space with ELVs as backup—must be reversed. ELVs must become the primary means of access, and redundantly so. The shuttle should serve only a secondary back-up role, and a contingent one at that. The reason for shuttle contingency status is that an important component of shuttle system reliability is stability in its manifesting. As the Rogers Commission pointed out, frequent changes in shuttle manifesting create ripples through the entire future STS manifest, disrupt the logistical support system and degrade astronaut crew training.[16] Stability in manifesting though, means inflexibility: the shuttle system cannot adjust easily to an emergency situation in ELV availability without threatening its own reliability.

If ELVs must become the primary U.S. means of access to space, what kind of systems should be made available? At present the Reagan administration envisions a mixed force of shuttle-sized ELVs, a new MLV, the stock of TITAN II ICBMs refurbished as ELVs, and the new heavy launch vehicle. The recent decision to procure an ALS is particulary important, for two reasons.

First, given the range of payload weights that constitutes launch demand in the 1990s, the present mix of ELV throwweight capabilities is too narrow. Second, almost without exception U.S. ELV technology is premised on 1960s concepts which are approaching technical limits and promise no fundamental improvement in cost per pound delivered to low earth or geosynchronous orbit.

The size and weight of communications satellites have been growing steadily since 1965. Intelsat VI satellites, for example, weigh more than 4,000 lbs. when in geosynchronous orbit, and some Defense Department geosynchronous satellites weigh more than 10,000 lbs. in orbit. Most U.S. ELVs cannot lift the heavier satellites to geosynchronous orbit, as Table 6 indicates.

Atlas, Delta and Titan vehicles are all derivatives of booster technology developed in the 1960s and upgraded during the 1970s. Although they represent different lift capabilities, all three nearly have reached limits of growth without fundamental redesign. None of them can compete with the Ariane 5, and only Titan can compete with Ariane 4 and the Japanese H2. The United States has only just begun to develop and acquire an unmanned ALS capable of delivering payloads to low earth orbit in the range 75,000—200,000 lbs. (1.5— 4 times the capacity of the shuttle). This means that the current ELV fleet will become increasingly capacity constrained and the United States will not have lift capabilities that will support interplanetary

TABLE 6
SATELLITE WEIGHT VS. ELV CAPABILITY

	Satellite Weight in Geosynchronous Orbit			
	1,500 lb	2,500 lb	4,000 lb	10,000 lb +
	Hughes 376 RCA 3000	Hughes 393 RCA 4000, 5000	Intelsat VI	DoD
Launch Vehicle				
Shuttle	Yes	Yes	Yes	No*
Titan IV	Yes	Yes	Yes	No**
Titan III	Yes	Yes	No	No
Atlas-Centaur	Yes	No	No	No
Delta	Yes	No	No	No
Ariane	Yes	Yes	Yes	?
Long March	Yes	Yes	No	No
Japanese H2	Yes	Yes	Yes	No

*Cancellation of the Shuttle/Centaur removed this capability.

**This mission sometimes can be met at the cost of lower station-keeping fuel loading and thus shorter satellite life.

Source: Author's estimates based on manufacturer's specifications

science missions, the space station and Defense (including perhaps SDI) until the late 1990s.

The Reagan administration has not yet made clear what size and composition of the U.S. ELV industry it expects in response to its ELV procurement and commercialization policies. Procurement of Titan IV ELVs assures that Martin Marietta will be in a position to offer launch vehicles (or launch services—see below) with its Titan III and IV vehicles—roughly at the mid-range of payload requirements. The Delta II MLV procurement is sized to provide launches for NAVSTAR global positioning system satellites. In payload terms these are roughly equivalent to the existing Hughes 376/RCA 3000 class of geosynchronous satellites—the low end of the market.

General Dynamics is offering commercial launches on its Atlas Centaur and has secured at least one commercial customer. In addition, NASA has announced that it will contract with General Dynamics to provide launch service for 5 payloads beginning in fiscal 1989.[17]

What should be the U.S. ELV industry objectives? First, we need redundant ELV access to space for critical national security missions. This means a minimum of two U.S. industry groupings, based on technologies sufficiently different that a failure in one will not require a standdown in the other. Given the emerging competition in commercial launch services, the United States reasonably could not expect to field more than three commecial ELV suppliers and probably only two over the long run. A decision to encourage the three U.S. ELV suppliers would imply a U.S effort to provide redundancy in launch services at the low end of the market (where the competition is heaviest) as well as in the middle (Martin Marietta's Titan).*

A decision to encourage two U.S. ELV suppliers would be an acknowledgement that a structure of two strong U. S. competitors is more robust than one of three weak competitors. But the important issue is that at least two U.S. ELV suppliers are necessary to provide redundant ELV access to space. Moreover, the existence of two or more U.S. ELV suppliers would convey the benefits of competition to government procurement—an increasingly important public policy objective.

The U.S. needs to take three further steps to insure the success of ELV commercialization policies, given the strategic framework outlined above. First, the present shuttle manifest should be revised to offload as many Defense Department and NASA ELV-class payloads as possible. Offloading the shuttle's government manifest would publicly demonstrate that the administration is serious about its commercialization policies, provide a government base of launch demand that would encourage private investment in U. S. launch services, and settle the shuttle launch price question decisively.**

Second, there should be a revision of government budget practices to permit advanced funding of *launch services* rather than advanced funding for government *procurement of launch vehicles for its own use*. The current Air Force MLV procurement illustrates the problem. In budgetary terms the Air Force can provide multiyear advance funding only through its procurement accounts—which means it must buy launch vehicles and manage its own launch operations. In other words the Department of Defense, given current government budgetary procedures, cannot easily shift from buying launch vehicles to buying launch services on a competitive price basis—in theory the principal feature of the administration's commercialization program. Moreover, government purchase of launch services would stimulate increased demand for launch insurance, which at present is purchased only for commercial satellite launches. Higher demand for launch insurance would strengthen the space insurance industry, which has

---

*It is doubtful that the U. S. would comtemplate paying for redundancy in HLV capability for the foreseeable future unless a decision is made to deploy SDI on a large scale.

**As long as the shuttle launches ELV-class payloads routinely, its price relative to that of ELVs will remain a thorny issue.

suffered in recent years from, among other things, its small business base.

Third, the U.S. must find a way to limit commerical liability for damage to government facilities caused by an ELV launch accident. A reasonable limit would be $100 million. Insurance could be procured for that liability at reasonable cost and would make it more likely that U.S. ELV companies could compete with foreign ELV suppliers.

Finally, there should be consideration of remanifesting commercial satellites on the shuttle between now and 1990. There are two reasons for doing so. First, many U.S. satellite operations are under contract to provide satellite communications services at prices based on low shuttle launch costs and resulting insurance prices. Offloading of government ELV-class shuttle payloads would permit the shuttle program to honor its commercial contracts. A revised shuttle manifest policy would weaken the short-term monopoly supplier position of Arianespace, which now commands 50 percent of the foreign and commerical launch market. The current forced migration of U.S. satellite operators to Arianespace and other foreign ELV suppliers will make it more difficult for U.S. ELV suppliers to compete once they enter the business.

To summarize: Current U.S. policy initiatives to restore American access to space are fundamentally sound; they need to be carried several steps further:

- The U.S. government should seek domestic capacity for assured access to space through vigorous encouragement of a redundant U. S. ELV industry comprised of two or three ELV suppliers;
- The space shuttle should be restricted to manned space missions;
- The U.S. government's budgeting procedures for space transportation should shift from procurement of launch vehicles, which places the government in the business of competing with other launch service suppliers, to purchase of launch services based on price competition;
- The U.S. government should set reasonable limits—about $100 million—on commercial liability for launch damage to federal property;
- Because U.S. government launch demand—whether for launch vehicles or launch services—is so large relative to the global market, it is important that government support of ELV services be carefully designed to insure growth in ELV payload capabilities; and
- The shuttle commercial manifest in the period 1988–90 should

be constructed so that U.S. satellite operators—those that survive*—
are not encouraged to establish the kind of long-term technical and
commercial relationships with foreign ELV suppliers that would make
it difficult for U. S. ELV manufacturers to enter the market in 1990.

Contemporary U.S. space transportation policy questions rank in
importance to the decisions President John Kennedy made when he
set the course for the Apollo program and captured the imagination
of the American people in the early 1960s. Kennedy also established
the Communication Satellite Corporation and thereby ushered in the
era of commerical satellite communications—now a multibillion-dol-
lar business that has revolutionized global communications.

Twenty-five-years later the U.S. finds itself faced with equally critical
decisions about American space leadership. If there is a difference it
is that the commercial prospects for use of space technology—not
withstanding the problems of 1986—loom very large indeed for the
next century. The Reagan administration and the Congress have
before them important opportunities to insure U. S. space transportation competitiveness for the rest of the century.

NOTES

1. It is not yet completely clear how the orbiter will be paid for or when it
is expected to become operational.

2. These two decisions flow from the Commercialization of Space Act of
1984 and the executive branch policy directive, National Security Decision
Directive 94, both of which preceded the Challenger accident.

3. *Report of the Presidential Commission on the Space Shuttle Challenger Accident*,
Vol. II, p. J–31.

4. *Pricing Options for the Space Shuttle*, Congressional Budget Office, 1985,
p. 16 (non-operational expenditures FY 1972–82 converted to FY 1986
dollars).

5. The U.S. Air Force partially succeeded in overturning this policy in
1984 when it secured funds to procure 10 Titan IV ELVs as a "complement"
to the shuttle.

6. Author's estimate.

7. "Post 51-L Assessment of Space Shuttle Flight Rates and Utilization,"
Jerry J. Fitts, NASA, July 21, 1986.

8. *Pricing Options for the Space Shuttle, op. cit.*, p. 4.

9. Author's estimate.

10. *Setting Space Transportation Policy for the 1990s*, Congressional Budget
Office, October 1986, pp. 15–17.

*Hughes Aircraft Company is a manufacturer and operator of communication
satellites and expects to survive.

11. *Aerospace Daily,* December 31, 1985, p. 311.

12. *Post-Challenger Assessment of Space Shuttle Flight Rates and Utilization,* National Research Council, 1986, pp. 7–9.

13. *Ibid,* pp. 30–31.

14. *Setting Space Transportation Policy for the 1990s, op. cit.,* p. 26.

15. *Report of the Presidential Commission on the Space Shuttle Challenger Accident, op. cit.,* Vol. II, pp. F1–5.

16. *Ibid,* Vol II, pp. J33–J43.

17. *Wall St. Journal,* October 20, 1987.

INTERNATIONAL COOPERATION AND COMPETITION IN SPACE: CHALLENGES FOR THE UNITED STATES

Marcia S. Smith

INSTEAD OF SPENDING massive amounts of money to build the "Star Wars" ballistic missile defense system, the United States could be mounting a joint U.S./U.S.S.R. manned expedition to Mars, to advance technology while demonstrating how the two countries can work together peacefully.

This, at least, is the premise of renowned astronomer Carl Sagan. Traditionally an opponent of sending people to other planets, in the belief that automated spacecraft can explore the Solar System as effectively and at less cost, Sagan nevertheless is hard at work promoting this idea. He is a strong critic of Star Wars (officially called the Strategic Defense Initiative), arguing that it should be terminated. Sagan and his supporters believe that one way to reduce the risk of nuclear war is to improve relations between the superpowers overall, and they suggest demonstrating that theme by embarking on a major, long-term cooperative space program—sending a joint American/ Soviet crew to Mars.[1] Whether one agrees with this premise or not, it is an example of the allure of proposals for cooperation in space.

Throughout the Space Age, the world has focused on the activities of the United States and the Soviet Union, but today, virtually every country in the world uses space in some manner, and many countries are developing space technology. The use of satellites for communications, weather observation and sensing of natural resources has become so commonplace as to be taken for granted. China, India, Japan and the European Space Agency (ESA; it has 13 European

The views expressed here do not necessarily represent those of the Congressional Research Service or of the Library of Congress.

countries as members) all have the capability to launch satellites into orbit. Other countries, such as Indonesia, Brazil and Mexico, have their own communications satellites. The number of players in space is growing, and with that growth comes more opportunity for cooperation as well as greater competitive challenges.

The benefits of individual national space programs include national prestige as well as military and economic advantages. Since 1957, when it launched the world's first satellite, the Soviet Union has been the chief competitor to the United States in terms of prestige. Currently, ESA and several of its individual members lead in developing the economic potential of space, and Japan's activities are increasing. So are China's. Several companies, including Western Union, have signed agreements to use Chinese launch vehicles to place their satellites in space.

The military aspects of space have received considerable attention since President Reagan inaugurated the Strategic Defense Initiative (SDI) in 1983. Military uses of space are not new, however. Both the United States and the Soviet Union have used space for military purposes since the beginning of their space programs. The superpowers are still the only important participants in the military space arena, but now that other countries have begun to build and launch satellites, this equation may change. The SDI program itself is international; agreements have been signed with several countries to participate in SDI research.

No country now has weapons based in space, but the Soviet Union has a ground-based antisatellite system, and the United States is developing an air-launched system. In the near future, at least, it is not likely that other countries will develop their own space weapons, though any country that can launch satellites has some potential to do so.

In areas of military activity that do not involve weapons directly, however, other countries are planning systems. Reconnaissance satellites provide detailed information on the military capabilities of countries throughout the world. This information historically has been exclusively the property of the U.S. and the U.S.S.R., but the French have announced plans to develop a reconnaissance satellite, probably in conjunction with other European countries. France already has a civilian remote-sensing satellite, the design of which can serve as the basis for a reconnaissance system. China and India also are thought to be developing reconnaissance satellites.

The ramifications this will have are unclear, but there are sure to be some. No longer will Western allies have to rely on information only from the United States. They will have their own data, from which they can draw their own conclusions about the world situation. Thus, even with respect to military uses, the international dimensions of space policy are changing. Further, although U.S. military and civilian space programs have been separate since the beginning, policies respecting these two uses are becoming increasingly linked. The terms cooperation and competition generally have been thought of as applying to civilian space activities, but the distinction is not as easy to make any more.

THE UNITED STATES AND THE SOVIET UNION: STILL THE BIGGEST GAMES IN TOWN

President John F. Kennedy's 1961 decision to initiate a manned mission to the Moon was made in response to the decaying international image of the United States, a decay caused in part by high-visibility Soviet space achievements. On May 25 of that year, Kennedy appeared before a joint session of Congress and called upon the nation to commit itself to the goal of landing a man on the Moon before the end of the decade.

This historic public address to the nation came only six weeks after Soviet cosmonaut Yuriy Gagarin became the first human to orbit his home planet, a year and a half after the Soviets had returned the first pictures of the far side of the Moon using an automated probe and three and a half years after Sputnik galvanized the world's attention.

The announcement of the manned [2] Moon program also came only five weeks after the Bay of Pigs fiasco had deeply embarrassed the new Kennedy administration both at home and abroad, and about a year after the U-2 spy plane incident in which pilot Gary Francis Powers was shot down over the Soviet Union had caused similar embarrassment for the Eisenhower administration.

Kennedy's decision posed quite a challenge considering the dearth of experience in accomplishing such a task—the speech came only three weeks after Alan Shepard became the first American to reach space, and he had not even gone into orbit.

Nevertheless, the country accepted Kennedy's goal. The reasons were both simple and complex. Some were concerned about national security—would the Soviets use the Moon as a military base if they got there first? Others worried about America's image abroad—was America second-rate in terms of technology and leadership? Kennedy had wanted a goal that was at once near-term and one at which the United States had a good chance of beating the Russians, to restore confidence in America. Heavily influenced by Lyndon Johnson, who had been intimately involved in space issues while in the Senate, and who as Vice President chaired the National Aeronautics and Space Council, the President made the choice to land a man on the Moon by the end of the decade. If it was unlikely that America could do it in nine years, it was just as unlikely that the Soviets could. Thus, competition for national prestige and national security were the driving forces behind the Apollo program.

Although Apollo is usually viewed as having been a nationalistic U.S. program, in a sense it was international. It fulfilled the dreams of people around the globe by demonstrating that humanity could leave Planet Earth and explore the new frontier of space first hand. The entire world, not just America, watched on July 20, 1969 when Neil Armstrong and Buzz Aldrin stepped onto the lunar surface. Scientists from many countries sought and received the opportunity to study the samples returned by the Apollo crews. America had won, but the world was truly the victor.

Aftermath of the Moon

The Soviets lost the race to the Moon, while Americans quickly lost interest in space. Attention turned back to the Vietnam War, the Great Society programs begun by President Johnson, and the resulting strain on the budget. Of the remaining seven visits to the Moon (1970–1972), the only one that attracted great interest was the near-tragedy of the Apollo 13 mission in which the crew was saved only by their own efforts and that of the ground support crew.[3] Funding for space activities dropped. The last three lunar landing missions were cancelled. The Apollo Applications Program, designed to run simultaneously with the lunar landings, but concentrate on using crews on space stations, was sharply scaled back to a single space station, Skylab.

It seems almost forgotten today, but the United States did have a space station once. The Skylab program (1973–1974) used Apollo capsules to transport three successive crews to dock with the space station. These missions garnered even less interest from the taxpayers than the last Apollo flights. Despite important scientific and operational achievements of the three crews, the public was so disinterested that when the second crew returned to Earth, complaints were heard across the country that the networks had cut into daily television fare to show the splashdown. When the final crew came back to Earth, there was no network coverage, even though the crew had set a new endurance record of 84 days.

Following Skylab, only one more mission remained in the Apollo series, and it did receive a lot of attention. The Apollo Soyuz Test

Project (ASTP), in which an American Apollo crew docked in orbit for two days of joint experiments with a Soviet Soyuz crew, was a blazing finale for the Apollo program started by President Kennedy. Ironically, in his last appearance at the United Nations, Kennedy had suggested that the journey to the Moon itself might be done cooperatively with the Soviet Union. Although it had begun as a race between the two countries, Kennedy eventually came to see the Apollo program in a new light. President Johnson in turn renewed the offer after taking office, but Congress and the Soviets themselves did not agree, and each country proceeded alone.[4]

ASTP was notable not only because it involved cooperation with the Soviet Union, but in a sadder sense, it was the end of an era in U.S. space exploration. No more Apollos, no more Saturn V launch vehicles. The United States decided to put manned space activities on hold until the space shuttle was ready, a six-year hiatus.

On April 12, 1981, the first shuttle lifted off from Cape Canaveral, coincidentally on the 20th anniversary of Yuriy Gagarin's space flight.[5] From 1981 through 1985, the shuttle enjoyed many successes, and brought public interest and attention back to the space program. But on January 28, 1986, the space shuttle Challenger exploded during launch, killing all seven crew members. This was only the first—although the worst—in a series of launch vehicle failures to occur over the course of the year.

Three months later, a Titan launch vehicle exploded during launch for the Defense Department. Three weeks after that, a Delta launch vehicle suffered the same fate. This left the United States with only one major launch vehicle, the Atlas, to take satellites into orbit. Since some of the Atlas components were similar to those of the Delta, however, it was grounded too. By the end of the year both the Delta and Atlas were back in service, but the stunning blow to a space program of which Americans had grown increasingly proud had left its mark with a dejected NASA and skeptical populace.

In the spring of 1986, *Jane's Spaceflight Directory*, an authoritative British journal, commented that the Soviets were 10 years ahead of the United States in space—an "almost frightening" lead.[6] The statement, coming as it did shortly after the Challenger tragedy, prompted major media headlines and many people incorrectly connected the two. After years of ignoring the Soviet space program, suddenly Americans again became interested in the "race in space." The fever-pitched whine of the Sputnik era began to rise in the throats of space enthusiasts across the country as they asked: are we behind?

The Soviet Union: Slow but Steady

In truth, at this point in the development of the two superpower programs, the Soviets are ahead in one segment of space activities—the use of crews in Earth orbit—and they were ahead in this area long before the Challenger tragedy.

The Soviets, having lost the Moon race because they were not able to develop a launch vehicle as powerful as the Saturn V, turned their attention to Earth orbit. In 1971, they launched the world's first space station, Salyut 1. Tragedy and technical setbacks in the early 1970s, however, slowed their manned program.[7] By the time the U.S. Skylab program ended, the United States had accumulated twice the amount of time spent in space by Soviet crews. One major difference, however, was that the Soviets never lost interest in space. Despite the setbacks, their program continued, and by the time of ASTP, two more Soviet space stations had been successfully flown.

Following ASTP, instead of placing manned space activities on hold, the Soviets redoubled their efforts. In 1977, the first second-generation space station, Salyut 6, was launched. With its two docking ports, the Soviets could now send resupply missions to the space station, enabling crews to stay for increasingly longer periods of time. Scientific experiments in areas similar to those explored by the Skylab crews could be conducted on Salyut for months. By the end of Salyut 6's lifetime, the duration for a crew in orbit had been extended to 185 days. This provided excellent opportunities for conducting research on materials processing in space, where new and better alloys, crystals, and pharmaceuticals can be produced because of the absence of gravity. Biological experiments could be conducted both on human reactions to long durations in weightlessness, and on growing plants, a necessity for closed ecology life support which will be needed if crews are to travel to distant places like Mars. In short, the Soviets had a laboratory in space providing a base for a flexible research program. Salyut 6 was replaced by its close cousin, Salyut 7, in 1982, and Salyut 7 is still in orbit (although it is not expected to be occupied again).

In February 1986, a third-generation space station with six docking ports, Mir (Peace), was launched. Mir increases the flexibility of the Soviet space research program by permitting the Soviets to outfit modules with special equipment and dock them with the station for varying periods of time. One module might be devoted to astronomy and another to materials processing, for example. Both of these applications may be performed better without crews, since they might be adversely affected by vibrations from human movements. A space station crew could ensure that everything was in working order on a module, then detach it from the space station. After a period of weeks or months, the module could return to the space station so the crew could change the experiments.

The Current Balance

Through their space station program, the Soviets have developed an impressive capability in low Earth orbit (LEO). In addition to the experiments which someday might result in profitable products for use on Earth, they have done wholesale repair of the space stations and ferried a crew between Salyut 7 and Mir. Although the first Skylab crew also performed heroic repairs, it was a long time ago (1973), and the work was not of the magnitude of the repairs on the Salyut stations.

The U.S. decision not to launch manned missions from ASTP until the first shuttle flight in 1981—not the Challenger tragedy—gave the Soviets the lead they now command in this area.

Shadowed now by the realities of the risks of spaceflight, the U.S. shuttle has nonetheless proved its worth as a space transportation system, repair shop, and scientific laboratory. The U.S. shuttle remains a technological triumph that no other country has yet equalled, although the Soviets are currently developing one.

In the article from *Jane's* cited above, the statement about the Soviet lead was, in fact, limited to the concept of the "practical utilisation" of space.[8] In terms of space science activities, however (such as operating observatories in orbit or sending probes to other planets), space applications (communications, weather, remote sensing, etc.) and basic space technology, the United States remains clearly ahead.

U.S. spacecraft have visited six of the other planets in the Solar System (Mercury, Venus, Mars, Jupiter, Saturn and Uranus), and one is now enroute to a seventh (Neptune). Automated probes preceded the Apollo astronauts to the Moon. The Soviets have sent automated probes only to the Moon, Venus, Mars, and Halley's Comet, and the Mars probes have been largely failures. The U.S. also has had several series of very successful observatories for studying the universe at different wavelengths, and analyzing the Earth's interaction with the Sun. Although the Soviets have also launched scientific satellites, no major discoveries have been made based on their data.

We should not be complacent in this area, though—the Soviets have invigorated their space science program. As will be noted below, other countries sent probes to Halley's Comet, too—the United States did not. The Soviets are planning a fascinating mission to Mars and its moons Phobos and Deimos in 1988 which involves significant cooperation with the West. Solar System exploration appears to be an area on which the Soviets intend to place a great deal of emphasis. We are leading now, but this could change in the next decade.

In the area of space applications, the United States is still in the lead. We pioneered the development of communications, weather, and land remote sensing satellites. The Soviets also have satellites for these purposes, but in the land remote sensing area, for example, they reply primarily on orbiting space station crews to obtain data using film cameras, rather than on automated satellites constantly transmitting data in digital form such as those used by the United States. That the United States leads in these areas is particularly important when discussing military space capabilities. Military space missions today, as in the past, are primarily in support roles— communications, navigation, reconnaissance, and weather. Like their civilian counterparts, U.S. satellites in all these areas are generally more advanced than Soviet spacecraft because of the technology difference.

Even in the area of sending people into space, the United States has the lead technologically. Although the U.S. shuttle is under repair now, it did work 24 times and will be restored to service. The Soviets are developing a space shuttle like ours and, reportedly, a smaller spaceplane, but neither is operational. The one area in which they are clearly in the lead is using crews on space stations. They were ahead in this area before the Challenger explosion, and they will be ahead even after the shuttle resumes service. Not until the United States has its own permanent space station will we be able to catch up with them in this one particular area.

Working Together

Now that attention is once more being paid to Soviet activities in space, the idea of cooperation has taken on new life, although, as noted, such concepts are not new. Agreements for cooperation in the 1960s and early 1970s eventually led to ASTP. The thesis was that the two countries could prove to the world that although they were still competitors in the arena of space, they could also work together to achieve common objectives.

As relations between the two countries in space over the following decade oscillated, the vulnerability of such cooperation to political tides became more apparent than ever. The five-year umbrella agreement that provided for joint working groups on topics such as exchange of biomedical data from spaceflights and future joint manned missions was renewed by President Jimmy Carter in 1977. Following the Soviet invasion of Afghanistan in 1979, however, U.S. participation in this work slowed considerably. When the agreement came up for renewal again in 1982, President Ronald Reagan refused to sign it because of what he saw as Soviet intervention in the affairs of Poland. Then, in August 1983, rumors abounded that the two countries were about to resume talks to reestablish space cooperation, but the Soviets shot down Korean Airlines Flight 007 and the effort withered on the vine.

Shortly thereafter, the Soviets coupled the issue of formal space cooperation with the elimination of SDI, arguing that they could not work hand in hand on civilian space goals with a country that was (in their view) single-handedly weaponizing outer space. In October 1985, a congressional delegation led by Representative Bill Nelson (D-Florida) tested the Moscow waters on space cooperation once again, but at the political level, SDI remained the dominant theme and major stumbling block.

Then, suddenly, in the summer of 1986, the Soviets decoupled space cooperation from SDI, setting the stage for renewed negotiations. An American delegation composed of NASA, State Department, Department of Defense and White House experts met with their counterparts in the Soviet Union in September 1986 and developed general language to re-create working groups. That these discussions occurred during the height of the Daniloff affair signalled the importance both sides placed on the subject. During the Reykjavik mini-summit, an announcement was made that agreement had been reached to cooperate in civilian space activities. Despite the rocky superpower relations following Reykjavik and the ensuing explusion of American and Soviet diplomats from each other's countries, further discussions were held in the United States and a final draft text resulted. U.S. Secretary of State George Shultz and Soviet Foreign Minister Eduard Shevarnadze signed it in early 1987.

It is apparent that both countries want cooperation at this point. Each gains in world opinion when there is cooperation in space—other countries like to see the superpowers working together toward common ends. The Soviets sometimes gain access to U.S. technology and management techniques. The United States gains insights into the secretive Soviet program, and in cases such as the exchange of biomedical data from spaceflights, information that we could not otherwise obtain. (The space shuttle, even when it resumes operations, will be able to stay in orbit only for 7 to 10 days. Soviet spaceflights have lasted as long as 237 days.)

So cooperation has its advantages, but it has disadvantages, too. For the United States, the largest concern is technology transfer. For the Soviet Union, it is the potential loss of face should the world discover that the United States has a substantial lead in the technology area. For cooperation to work, both concerns must be satisfied. Each country will cooperate only if cooperation is in its best interest. There is no single solution for U.S. technology transfer concerns; each case must be handled on an *ad hoc* basis. The Soviets will obviously cooperate only where it shows them in a favorable light.

When considering possible areas for cooperation, it would seem that the Soviets would probably prefer to avoid cooperation in the manned arena until they have their own shuttle. As impressive as their space stations are in terms of scientific potential, technologically they are not sophisticated and the sight of a sleek U.S. shuttle snuggling up to a "low-tech" Soviet space station probably would not paint the Soviets in the light they prefer. From a U.S. standpoint, although President Reagan suggested a joint shuttle/Salyut mission to the Soviets in 1984, there could be technology transfer concerns if such a program were adopted.

Space science seems to be the area on which most attention is being focused today. The United States can learn much using Soviet data, with little or no technology transfer, and the Soviets gain prestige when we have to go to them for information. Providing biomedical data from the space station crews would be acceptable—the Soviets have a lead in this area and the U.S. cannot obtain data itself for such long durations. Exchange of data from planetary probes would be acceptable also—the Soviets have a growing program for sending probes to Mars and other bodies in the Solar System which relies increasingly on cooperation with countries outside the Soviet bloc.

The ability to compare data from various spacecraft studying the same object can increase scientific knowledge several fold. For example, the Soviet Union will send a probe to Phobos, one of the moons of Mars, in 1988, a mission that is enormously appealing to the world scientific community. Many Western countries are cooperating with the Soviets on this mission. The United States will send the Mars Observer in 1992 and it will focus on studying the planet itself; data from both the U.S. and Soviet probes can be used to gain new insights into Mars and its moons.

Someday, the two superpowers may work together to design probes to study Mars in more detail. The head of the Soviet Institute for Space Research, Dr. Roald Sagdayev, has informally proposed that the United States send a vehicle to Mars that could rove across the surface, collecting samples from different locations. The Soviets would send a separate spacecraft capable of returning those samples to the vicinity of Earth for study. The samples could not be returned directly to Earth for fear of contamination. Leonard David, a private consultant in the United States, has suggested that the two countries first work together to build an orbiting quarantine facility for receipt of the samples. This could demonstrate that the two countries could work together on a long-term project and provide a facility for studying the samples.

The mission is an excellent candidate for cooperation from several viewpoints. First, almost no technology transfer issues would be involved, since the only place the two spacecraft must work together is on the surface of Mars. Second, if either country were to withdraw from the program for political, technical or budgetary reasons, the other probe would still have a viable scientific mission alone. If the mission did proceed as planned, however, the scientific knowledge gained would be tremendous.

Carl Sagan's proposal of sending U.S. and Soviet crews to Mars together would clearly entail a major undertaking. Perhaps these other activities could serve as steps toward his goal, although it might take longer than Sagan would prefer. Waiting for the rover/sample return mission and perhaps the construction of a space quarantine facility would require many years. In any case, prospects for U.S./Soviet cooperation in space, on the smaller level of joint working groups or developing automated spacecraft to visit Mars, look especially promising today. There are, of course, few areas of U.S. policy where changes occurs as quickly as in superpower relations, and some difficulties must be expected. Where both countries have something to gain and little to lose, however, cooperation can prevail.

THE BROADER INTERNATIONAL SCENE

Attention so far has focused on cooperation and competition between the U.S. and Soviet Union, but they are no longer the only players. International cooperation and competition have become much broader as the years have progressed.

The law that created NASA in 1958 included language stating that NASA "may engage in a program of international cooperation" in its activities.[9] Four years later, when Congress passed the Communications Satellite Act of 1962 and created the Communications Satellite Corporation (COMSAT), it declared U.S. policy as establishment of a global commercial communications satellite network "in conjunction and in cooperation with other countries, as expeditiously as practicable."[10] In 1964, the International Telecommunication Satellite Organization, Intelsat, was established. It now has 112 members around the world.

The U.S. civilian space program has thus always had a strong international component. The United States has concluded over 1000 agreements with more than 100 countries. The cooperation entailed by these agreements has taken many forms: exchanging scientists, providing launch services in exchange for data received by the spacecraft, joint construction of scientific experiments and spacecraft, and the construction of a spacecraft for joint use by the countries involved.

The road to cooperation has not always been smooth. The annual budget cycles in the United States often complicate our participation in international projects since the United States cannot guarantee its funding for the project through to the end. Each year, every program faces the possibility of cancellation or deferral. To provide an overview of the types of international programs that have been conducted, and the problems they encountered, a few examples will be given here of some of the larger and more visible projects. It should not be forgotten, however, that many successful cooperative ventures have involved relatively low levels of effort, such as allowing other countries to study samples returned from the Moon by Apollo astronauts. These programs build good will and train people to work in space-related fields.

Europe

More than with any other group of nations, the United States has cooperated to the greatest extent with Europe, both individual countries and the European Space Agency (ESA). Created in 1975 as the merger of the European Space Research Organization, which developed scientific satellites, and the European Launcher Development Organization, which attempted to build a European launch vehicle, the agency itself is a testament to international cooperation, working out the competing and conflicting desires of its 13 member states.[11]

ESA has developed a launch vehicle called Ariane. Its operations have been turned over to a quasi-private organization called Arianespace, while ESA is now responsible only for developing new versions of the vehicle. Interestingly, one impetus for developing a European launcher was that the United States initially refused to launch a French/German communications satellite, Symphonie, in 1974 on the grounds that it might compete with Intelsat. This action helped convince the Europeans that they must have their own access to space. Now Ariane is the biggest competitor to U.S. launch vehicles in the lucrative market of launching satellites. Like the shuttle, Ariane failed early in 1986, with many months of down time before resumption of service in September 1987.

Although cooperation with the United States has been important to ESA and its member states, the Europeans have had several highly successful programs of their own. In 1986, for example, their space-craft Giotto was one of the international probes that intercepted Halley's Comet. It came the closest of the five probes (two Soviet, two Japanese, one European) and returned data of significant interest.

Among the many successful U.S./European cooperative programs is Spacelab, a scientific laboratory that flies in the cargo bay of the space shuttle. Spacelab was built by ESA under a cooperative agreement with NASA. Although now the Europeans are critical of the agreement they signed for the Spacelab program, at the time it was cheered as an opportunity for European industry to jump into the high technology world of space hardware. For all its problems, Space-lab did do what it was intended to do—gain entry for European industry into the space hardware business.

A less successful cooperative venture was the International Solar Polar Mission project. For this project it was agreed that the United States would build one probe and the Europeans another, both designed to study the Sun's poles, areas which cannot be observed from Earth. Both spacecraft would be launched and tracked by the United States, and we were to provide the special power sources (radioisotope thermal generators) required. The Reagan administration cancelled the U.S. probe because of budgetary considerations. Despite strong protests by ESA members to the State Department, and support for the program in Congress, the administration would not be moved. The United States agreed to launch the European satellite, provide tracking and data support, and the power source, but not a spacecraft. The Europeans, having invested great sums of money in their probe, unhappily went ahead with a one-spacecraft mission. The ESA probe, which came to be called Ulysses, would have been launched on the space shuttle in 1986, and is now scheduled for 1990.

Japan

Japan's space program has centered on cooperation with the United States since its beginning. The primary Japanese launch vehicle (the N) is an American booster, the Delta, manufactured in Japan under an agreement with the U.S. government that stipulates that Japan will use it only for peaceful purposes, and that Japan cannot use it to launch satellites for other countries without our permission. The Japanese are also constrained by agreements with fishermen who work off the coast of their main launch site on Tanegashima, providing that they can only launch from there twice a year. Thus, the Japanese are not now an economic competitor to us in the launch business.

The Japanese have another launch site, at Uchinoura, for the small Mu series of launch vehicles used for space science missions. Two Japanese probes launched by Mu rockets were among the international armada sent to meet the arrival of Halley's Comet. The Mu cannot lift heavy satellites into orbit, however, and fishing-related restrictions apply to Uchinoura as well. Japan is now looking at the possibility of building another launch site, possibly in cooperation with the United States and other Pacific Basin countries, perhaps on Christmas Island. They are building their own launch vehicle, the H series, and when the H II is available in the early 1990s, Japan will no longer be limited by U.S.-imposed restrictions on how they use their launch capability. If the market is there, they will almost certainly enter the competition for launching satellites into orbit for profit.

Many Japanese satellites, in fact, are built by U.S. companies, but Japan has insisted that its engineers participate in the construction so they can learn to build their own. They now have the most advanced communications satellites in the world.

Canada

Long an active participant in space through the communications satellite field, Canada was the second nation, after the Soviet Union, to have a domestic communications satellite system. Through contributions to the U.S. shuttle program, Canada has become a leading player in space despite the fact that it does not have a launch capability. Its aerospace companies compete with U.S. firms for building communications satellites throughout the world. The "Canadarm," or remote manipulator system on the space shuttle, has become well known for its yeoman work in retrieving satellites from open space. Because of its strong interest in space, Canada is now creating its own space agency.

The Space Station—An International Endeavor

In his 1984 State of the Union address, President Reagan announced that he was directing NASA to build a permanently occupied space station, and noted that it would be built with international participation.

The international participants—ESA, Japan and Canada—want to think of the space station as a truly international endeavor, not a U.S.

effort to which they are being added. Thus, the question of whether it is a U.S. space station, a U.S.-led space station, or an international space station has become an important issue. The distinctions may sound academic, but in terms of the negotiating postures of the countries concerned, it makes a considerable amount of difference.

The lessons of Spacelab and ISPM have not been lost on ESA and it is taking a strong position on the space station. The essence of the problem is that the United States wants to make sure that it is not giving away too much on a program that it initiated, while the Europeans want to be treated like equals and not ordered around by Americans who have not come to grips with the fact that the European space program has grown up. While some of the arguments may be disregarded as posturing, the fact is that this time the Europeans can decide that they would rather go it alone.

In 1986, ESA approved a plan to develop a reusable spacecraft to carry crews into orbit and return them to Earth. Called Hermes, this spaceplane would be launched on an upgraded version of Ariane and would make the Europeans autonomous in space. No longer would they be dependent on the United States to take them to space laboratories where they can conduct research on new materials that can be developed in microgravity. Although it is not known today what metal alloys, pharmaceuticals, or crystals may come from such research, the potential is great.

Japan and Canada, though less wary, also want their contributions to be appreciated. Japan will build a habitable research module, while Canada will create a system to serve the functions of the shuttle's "arm."

The resolution of these difficulties in cooperating with our "traditional" partners in space will not be easy, but is critical to assuring U.S. preeminence in space and the technology it brings.

THE LESS DEVELOPED COUNTRIES

In this discussion of space cooperation, the less developed countries (LDCs) should not be omitted. In fact, two of them, China and India, have their own ability to launch satellites into space .

Although India's launching ability is rudimentary today, it has announced plans to build launch vehicles with greater capability over the next few years. The Indians have already built, or paid to have built for them, satellites for remote sensing, communications, and weather observations. They have decided that their launch site at Sriharikota is insufficient, and a second launch site will be built soon.

India often portrays itself as the leader in the Third World for space, but, in fact, China is far ahead at least in one area—the development of launch vehicles. The Chinese CZ-2 and CZ-3 boosters are being successfully marketed internationally. With the current problems with Western launch vehicles, China is making inroads into the launch services business, and has signed contracts with several Western companies, including Western Union. Between being able to give assurances of launch date opportunities and providing insurance at rates almost impossible to obtain in the West, China has much to offer for those willing to take a bit of a risk.[12]

The United States has had many cooperative programs with the LDCs, India and China included. Among the most visible is the Landsat program for land remote sensing. The United States was the first country to have a land remote sensing capability and it has been our policy to make the data available to all countries on a non-discriminatory basis for a nominal fee. We encouraged other countries to build their own ground stations for direct receipt of Landsat data, and created centers in the LDCs to teach methods of interpreting the data. With this information, countries can better monitor the existence and use of their own natural resources. Crop yields can be forecast. Areas where desertification is occurring can be identified. The program has been lauded far and wide for the data it provides and the openness of access to that data.

Landsat was developed by NASA and later operated by the National Oceanic and Atmospheric Administration. The Carter and Reagan administrations decided that the system was ready to be operated by the private sector, and after years of tortuous negotiations, a private company called EOSAT took control in 1985. Promised government subsidies to get it moving, EOSAT now finds itself in a difficult position between the Reagan administration (which has not requested the funds it promised), and Congress (which has restored the funds contingent upon certain reports from the administration). It is not inconceivable that EOSAT will fail and the United States, which pioneered the development of this technology and earned tremendous international goodwill through the Landsat program, will no longer have a system. The LDCs will have to fend for themselves in what is sure to be a highly competitive market for this type of data.

France has a remote sensing satellite system already—SPOT. It decided that remote sensing has a number of values, including profitability. The French created a quasi-private agency, Spot-Image, to market the data. ESA and the Japanese are both planning to launch land remote sensing satellites soon, and the Japanese launched an ocean-sensing satellite in 1987 to provide data that may pinpoint the location of the best fishing grounds. The Soviet Union has announced plans to market its remote sensing data. If the United States wants to give up its lead in this area, there are many competitors waiting to take over.

The United States pioneered communications satellite technology, too. Today, however, Japan and Italy in particular are leading in the development of new communications satellite technologies. The only civilian government program in the United States for developing such technologies, the Advanced Communications Technology Satellite program in NASA, faces funding battles each year. Although there are military programs for advancing communications satellite technology, many of the concerns (such as the ability of an adversary to intercept and/or decode the transmissions) are not the same as those of the civilian/commercial sector. Also, much of the research is classified and cannot be easily transferred to the commercial sector. There are international competitors waiting to take over in this area, too.

SPACE 2000—A POSSIBLE SCENARIO

To put the relative capabilities of other countries in perspective, it might be useful to consider a somewhat fanciful scenario of what space might look like some years hence if the United States slows its civilian space program.

Let's imagine that it is the year 2000.

The Soviet Union has three space stations accommodating 15-20 cosmonauts, and a multitude of modules in different orbits for performing tasks such as materials processing of new alloys and better crystals on an industrial level. Orbital transfer vehicles ferry crews back and forth between these modules, harvesting the products being created and changing scientific experiments. They have cornered the market on gallium arsenide crystals for computer chips, and anyone wanting a sixth-generation computer must rely on the Soviets for supplies. Having overcome their propensity for secrecy, they are also now major suppliers of land remote sensing data and launch vehicle services.

They are close to completing their first solar power satellite in GEO, a development helped by their concentration on building GEO platforms earlier in the decade. Some of the electricity produced will be beamed back to Earth, while more will go to power their new outpost on the Moon. Although still dejected that they did not beat the Americans to the lunar surface in the 1960s, they feel they can make up the difference by establishing a permanent presence there in the early 2000s. The raw lunar material they mine will be used for shielding Soviet space stations both in Earth and lunar orbit, where more lunar material will be refined for building additional solar power satellites to feed an ever energy-hungry Earth.

The automated probes sent to Mars in the 1980s and 1990s were great successes, and together with Europe and Japan, the Soviets will mount a manned mission to Mars in the next decade or so.

The Europeans, working with the Japanese and Canadians, have their own space station serviced by the Hermes vehicle. The Germans have created a new metal alloy in space that is of lighter weight and greater durability than the composities of the 1980s. Airplanes, ships, automobiles—all are now made with the German product. GEO platforms are serviced by MARS, Inc. (a consortium of Mitsubishi, Aerospatiale, and Spar), with customers such as the United States paying top dollar for fixed and mobile communications services, as well as navigation.

The Chinese, who captured a large percentage of world-wide commercial launches have begun construction of orbital transfer vehicles for transporting people and materials from low earth orbit to GEO and the Moon. They had once decided to build their own space shuttle, but later concluded that the more lucrative market was in on-orbit services.

The United States, relegated to space shuttle flights and the anachronistic Spacelab, is still performing basic research in many of the areas already earning profits for other countries. American ELVs can't compete with the government-subsidized launch vehicles elsewhere. Without a space station, no need was seen for developing U.S. orbital transfer vehicles and other on-orbit services. Having balanced the budget at the price of technological leadership in the world, the United States is facing another economic crisis with soaring unemployment rates, rising inflation, and trade deficits that make the 1980s look like the good old days.

What America Should Do

No one can argue with the good intent of attempting to balance the federal budget, but one must look at the impact on the United States as a whole of taking certain actions. The civilian space budget consumes less than one penny of every tax dollar. The military space budget costs twice that—a total of less than three cents.

This seems a small investment for maintaining U.S. competitiveness in an ever-changing, technologically-sophisticated world, while at the same time providing an outlet for the human yearning to explore new frontiers.

While funding is obviously crucial, an even more important ingredient is commitment to long-term goals. Proposals for a long-range space program were made by the National Commission on Space in

1986 in its report *Pioneering the Space Frontier.*[13] The Commission, established by Congress and the President, spent one year delineating objectives and goals for the U.S. civilian space program over the next 50 years. Its main recommendation was that the United States lead the way in opening the inner Solar System for science, exploration, and development, including the establishment of settlements on the Moon and Mars. The Commission did not envision the United States doing this alone, but in concert with the most creative minds throughout the world. The Commission's report is not national policy, however, and the report has been languishing in the Executive Office of the President since it was submitted in July 1986. It is this type of commitment that is needed if the United States is to remain a world leader in space.

Why is it important for the United States to lead in space? There are several reasons, not the least of which is competitiveness. Competitiveness has become a watchword in Washington, but few people seem to understand the connection between technological competitiveness and investment in high technology programs such as space. Other countries have no trouble drawing the connection; hence, there are space programs in countries with much less well-developed economies than the U.S., like China and India. The Soviet Union also understands that space programs contribute to a nation's technological status in the world. During a visit to the space launch facility at Baykonur (also called the Tyuratam facility) just before the Soviets tested their new, heavy-lift launch vehicle Energiya, General Secretary Gorbachev stated that Baykonur "has become a symbol of our homeland's greatest exploit—a triumph of Soviet science and the great potential of the socialist system." He continued:

. . . Everything here . . . has been produced by us in the U.S.S.R. It is all high quality and state of the art technology.

Once again a simple but very important question comes to mind: Why do we at times try to acquire even simple items from abroad if we are today capable of resolving such vast, large-scale and complex tasks? Everything seen here leads once again to the deep conviction: There is no reason for us to go abroad, hat in hand in this way. No embargoes, no ban imposed by certain foreign circles on selling us technology and equipment will slow down the development of our country or the implementation of the great social and economic plans connected with restructuring and the acceleration of our economy.

This once again convinces me of the need to give every support to our science, our scientific intelligentsia, engineers and designers.[14]

As for space cooperation, whether with the Soviet Union, LDCs, or other industrialized democracies, space projects of greater complexity can be undertaken if countries pool resources and talent, resulting in greater gains for all concerned. If international relations are improved at the same time, it is an added bonus.

Thus space competition and cooperation are each important in their own ways. To participate in either or both, the United States must have a firm commitment to a long-term space program. Space is the next frontier. Can America afford to be on the sidelines?

NOTES

1. This is a very abbreviated version of Sagan's arguments. See Carl Sagan, "Let's Go to Mars Together," *Parade,* February 2, 1986, pp. 4–7.

2. There are many, including the author, who would like to find an easy to use gender-neutral synonym for "manned" spaceflight. Throughout this article, great pains have been taken to refer to "crews" and "automated probes" to solve the problem of references to "manned" and "unmanned" programs. Nevertheless, "manned" sometimes seems unavoidable; the author apologizes.

3. Although the Apollo 13 crew survived, let it not be forgotten that three men—Virgil "Gus" Grissom, Ed White, and Roger Chaffee—gave their lives to the space program on January 27, 1967, when a fire erupted in their Apollo capsule during ground tests at Cape Canaveral before the first Apollo flight.

4. For a history of the Apollo program, see: House Committee on Science and Technology, *United States Civilian Space Programs 1958–1978: Vol. 1,* U.S. Government Printing Office, Washington, D.C., 1981, pp. 379–428. This report, coupled with Vol. 2 (1983), provides detailed historical information on all U.S. civilian space programs.

5. The shuttle mission was supposed to be launched on April 10, but was cancelled minutes before liftoff due to computer problems. Thus, its timing was purely accidental.

6. Reginald Turnill, ed., *Jane's Spaceflight Directory 1986,* Jane's Publishing Company Ltd., London, 1986, p. 17.

7. The Soviets have lost four cosmonauts in spaceflights. On April 24, 1967, three months after the Apollo fire, the pilot of the first Soyuz spacecraft was killed upon impact with the Earth when the Soyuz parachute lines became tangled during descent. In 1971, after the first successful mission on board a space station, the three-man Soyuz 11 crew was asphyxiated when a valve was not properly closed during descent. The spacecraft's atmosphere vented into space; the men were not wearing spacesuits.

8. Turnill, *op. cit.*

9. Sec. 205, National Aeronautics and Space Act of 1958, P.L. 85–568.

10. Sec. 102, Communications Satellite Act of 1962, P.L. 87–624.

11. ESA members are: Austria, Belgium, Britain, Denmark, France, Federal Republic of Germany, Ireland, Italy, the Netherlands, Norway, Spain,

Sweden, and Switzerland. ESA has a technical agreement for cooperation with Canada.

12. The CZ-3, capable of taking satellites to geostationary orbit (GEO), has been launched only three times, and the first time it failed. Geostationary orbit is a special orbit located over the equator at an altitude of 35,800 kilometers. Satellites placed there maintain a fixed position relative to points on the Earth. GEO is very useful for applications such as communications satellites.

13. U.S. National Commission on Space. *Pioneering the Space Frontier*. Bantam Books, New York, 1986.

14. "Gorbachev Speaks at Baykonur Cosmodrome in Leninsk," *Pravda*, May 14, 1987, pp. 1–2. Translation from FBIS, *Daily Report—Soviet Union*, May 14, 1987, pp. R 1–2.

OPPORTUNITIES FOR ASTRONOMICAL OBSERVATIONS FROM SPACE

Frank D. Drake

HOW MANY OF our precious resources—financial, technical, and human—should we invest in the capacity to conduct astronomical observations from space? This very important question, with its widespread implications for the future of scientific research, is one that seems very hard to address in an era of strained resources. Yet, the benefits to be gained are quite clear. The results of three decades of observations of the cosmos from spacecraft have shown the wealth of scientifically important, provocative, and unpredicted phenomena which can be observed only with the use of spacecraft. The investment of substantial resources is sure to produce remarkable results. Indeed, if we have learned anything, it is that the universe is a far richer and more complex place than the human mind can imagine; reality far outdistances in variety the most creative ideas of our greatest scholars.

There are a number of formal science policy committees in the U.S. involved in constructing overall plans for astronomical space research. From the evidence of achievement, it is clear that these committees have succeeded in maintaining a superb record in balancing the costs and benefits of specific programs. There is little one can say, then, which does not echo the work of those committees in some measure. However, any individuals engaged in a potential policy planning exercise are likely to be inhibited in their thinking, both knowingly and unknowingly, by their sense of what is possible in the foreseeable future. Their thinking is strongly influenced, as it should be, by their estimates of the resources likely to be made available to the program. Their charge is to construct a program which is reasonable in the context of the political and financial realities of the time.

But attitudes and priorities can change. The prospect of especially attractive enterprise can lead to massive new funding: witness the birth of the U.S. space program itself, and support for the Strategic

Defense Initiative. Space astronomy offers some remarkably tantalizing projects, as we shall see. But in many cases they require a level of
technological endeavor and financing which seems beyond (although
not by much) the resources we might expect from Congress. Such
projects are too often banished to the appendices of reports or to
chapters on the far distant future. Not so here. We will state what
would be within our immediate grasp in the best of all worlds.

Life, Landers and the Solar System

Throughout the Solar System, not in just one but in many places,
we can expect to see in action processes like those which gave rise to
life on earth. As living things, this information is of enormous interest
to us. We wish to know our origins; we wish to know whether we are
rare, or even unique. We wish to know if the curious and complex
chemistry which makes terrestrial life possible is a fluke, or the rule
for life in the universe. Are we an optimum form, or just one of many
forms which work about as well? Does chemical evolution converge on
"our" way as the best way, or are there many ways, perhaps even
better ways, to make a living, reproducing, mutating, thus evolving,
entity?

We already have conducted "fly-by" reconnaisance of all the planets
and their major satellite bodies, out to and including Uranus. The
findings: an enormous variety of form, chemistry, temperature and
physical states. We are surprised, but we shouldn't be. After all, the
earth is far more complex than anyone would guess from examining
it from a distance. One conclusion seems very clear: the complexity
of the phenomena that characterize solar system objects is very likely
to defy the ability of remote observations to provide a full understanding of those phenomena. On-site observations, observations
from landers, clearly are called for. With the exception of the Galileo
mission to Jupiter, no such mission is proceeding, yet there are a large
number of fascinating opportunities.

Europa. A satellite of Jupiter, Europa apparently is covered with a
mantle of water ice in which large fractures, extending many hundreds of kilometers, have occurred. Almost surely, below some depth
there is liquid water, possibly an ocean, with temperatures suitable
for life. What is the nature of this ocean? Are there undersea vents as
in terrestrial oceans? Is there a primitive biochemistry in this ocean,
and perhaps even entities which fit the definition of living things?
Designing experiments which could deal with these possibilities would
be a real challenge, as would be the mission itself.

It would probably be most productive to land in one of the surface fractures, but this runs contrary to our usual policies. When the Viking mission went to Mars, we went to great lengths to insure that the two spacecraft would land in the safest possible places. We succeeded, but in the process landed on the part of Mars which is least interesting and least likely to reveal to us the history of Mars. The landing site was equivalent to the most barren part of the Sahara desert here on earth. What would such a landing reveal about the richness that characterizes our planet? The upshot of this is that with Europa, as with many other Solar System bodies, if we are to realize the greatest gain, we will have to take risks. Potential costs might be very high, including the possible loss of spacecraft, but the overall return would be greatly enhanced. It would be a challenge for NASA to take this daring approach and to convince Congress that it is, after all, the most prudent way to get the greatest return for our efforts, and that the outcome would be well worth it.

Ganymede. Another of the giant satellites of Jupiter, Ganymede shows a landform found nowhere else in the Solar System. Systems of giant furrows or ridges in the ground, resembling an orderly plowed field, appear across the surface of Ganymede, with the furrows tens of kilometers apart. What is the origin of such a terrain?

Io. The smallest of the Galilean satellites of Jupiter, Io is one with continuous, intense volcanic activity. The volcanoes spew a material rich in sulphur. And what else? There must be a very bizarre chemistry at work here.

Titan. The largest satellite of the Solar System, Titan is the only one with an atmosphere—an atmosphere we know to be rich in nitrogen and hydrocarbons. There is evidence for the presence of an enormous number of organic molecules, including, for example, hydrazine, which has only recently been found in the laboratory to be an excellent precursor to the development of many of the molecules of terrestrial life. Presumably beneath the clouds of Titan there is an extensive ocean of ammonia or some other solvent, in which are dissolved the products of eons of chemical reactions in the atmosphere. What has this produced? To find out, we must deposit not a lander, but a floater, in this ocean.

Iapetus. A satellite of Saturn. About half of the surface of Iapetus is evidently ice covered, and is one of the whitest surfaces in the solar system, but the remainder is nearly black, probably the darkest surface we know of in the solar system. The little light which is reflected from the dark part has a spectrum close to that of the material found in the carbonaceous chondrites, a rare form of meteorite rich in carbon, and containing in small amounts most of the molecules necessary to make the living things of earth. What is the nature of this remarkable deposit? Where did it come from?

Miranda. A satellite of Uranus, Miranda shows a remarkable variety of landforms, almost as though it is the result of combining a number of smaller bodies into one. One of those landforms resembles the peculiar "plowed" landform of Ganymede. What is the nature of the boundaries between these landforms? These boundaries could well reveal the evolutionary history of the body.

Triton. A satellite of Neptune. There is good evidence that on Triton there is an ocean of liquid nitrogen, or "liquid air." What is the chemistry of that ocean? What is there to find on the shores of an ocean of liquid nitrogen?

A comet, any comet. Comets are almost surely the residue of the material from which the solar system was formed. However, they have spent almost all of their time in the deep recesses of the solar system, a place of darkness and incredible cold. As a result, their basic materials have remained unaltered over the 5 billion year history of the solar system. Samples from the interior of a comet will reveal the raw materials from which the solar system was born. These materials will be very close to the materials from which all stars have been born since about the time of the sun's birth.

Samples from the exterior of a comet will reveal the consequences of billions of years of exposure to the sources of energy which exist far out in the Solar System—for example, the faint light of the sun, including ultraviolet light; bombardment by solar wind particles, mainly hydrogen atoms; and bombardment by cosmic ray particles. It will be interesting to see if, in particular, the molecules relevant to the formation of life as we know it are formed in such a milieu.

Asteroids. Landings on asteroids would reveal much about processes in the solar system. With on-site observations on the large asteroids, such as Ceres, we also will again be able to study pristine bodies whose surfaces have been subjected to bombardment by small bodies, faint sunlight, solar wind particles, and cosmic rays. Will we see a duplication of the materials found in some meteorites? In the smaller asteroids we see the remnants of objects which have been fragmented by catastrophic collisions. Just as the study of the sediments of the earth reveal the history of geological processes on the earth, the study of these asteroids should reveal the long history of events in the solar system.

Lastly, probes into the major planets, similar to the Galileo probes already planned, would reveal details both of the major physical processes and the very rich chemistry which form the atmospheres of these planets. These atmospheres can aptly be deemed "cauldrons," for we have seen directly the enormous convection and turbulence which occurs there, moving masses of gas among regions of greatly differing temperatures, pressures, and chemical make-up. The variety of colors seen in these atmospheres, and the results of laboratory experiments which have produced products of similar color, organic materials of very high molecular weight, suggest strongly that we will find that these planetary cauldrons have produced a cornucopia of molecules, particularly organic molecules. It would be fascinating to see how far these processes have gone towards establishing a biochemistry.

In the missions alluded to above, there of course would be other studies of interest. Magnetic fields. Atmospheric structure. Winds. Weather. Ionospheres. Where present, planetary topography. All of these, observed in and near the planet, would reveal far more than can be revealed by the remote sensing provided by fly-by or orbiter spacecraft. A lander mission is a difficult and expensive one. But our understanding, as well as our direct experience with landers, show us that there is really no substitute for them with regard to many key questions about the Solar System.

For many years, NASA advertised almost every mission to a planet as a mission which could prove the existence or absence of life on the planet. This reflected the correct perception that the world at large has a real hunger to know the meaning, the basis, the significance of life on earth and elsewhere in the universe. NASA derived a great deal of support from this hunger on the part of humanity. But the truth was that none of these missions, except for Viking, could have proved the existence of life anywhere. NASA was overselling. In time, the public and Congress recognized this, and it contributed to diminished support for NASA, as well as a degree of cynicism. But the hunger is still there, in very large measure, as is easily established by observing the public interest in movies, books, and television programs on life in the universe. Now a specious argument has become a genuine one. If we exploit our abilities to achieve landings, NASA can do what it has long been expected to do.

Beyond the Solar System

There are equally great opportunities for space-borne instruments to make major discoveries in deep space, in the realm of the universe beyond the Solar System. The importance of the capability of spacecraft to observe at wavelengths which do not penetrate the atmosphere, such as the ultraviolet, X-ray, and many infra-red wavelenghts, is not arguable. The ability to make superb observations at these wavelengths has already been demonstrated by such spacecraft as Einstein, the International Ultra-violet Explorer (IUE), and the Infra-red Astronomy Satellite (IRAS). There is clearly an important place for such instruments, of ever increasing sensitivity, in our space program.

The next step in the development of our deep space observation capacity will be the launch of the Hubble Space Telescope (HST), scheduled for late 1988. This 2.4-meter telescope of conventional design will be especially powerful at ultra-violet wavelengths which do not penetrate the atmosphere. Despite its relatively small size, it will provide spatial resolution which excels that of current ground-based telescopes, because the sharpness of its images will not be deteriorated by atmospheric turbulence. It also will be looking at a darker sky because of the absence of atmospheric air glow over it. This, plus its sharper imaging capability, will allow it to see much fainter objects than are detectable from the surface of the earth. However, resolutions comparable to those of HST also may be achieved by new, giant, ground-based telescopes such as the Keck 10-meter telescope, a joint project of the University of California and the California Institute of Technology. The achievement of such resolutions depends on the application of sophisticated devices which act on incoming light waves so as to correct, or actively control, the distortions introduced by atmospheric turbulence.

A realistic estimate of the resolution which might be achieved, either by HST or by actively-controlled ground-based telescopes, is roughly 50 thousandths of an angular arc-second, or 50 milliarcse-conds. This is ten to twenty times better than the resolutions achieved in the best conditions by the best contemporary instruments, and will allow major advances in astronomical research. Nevertheless, the limitations on our ability to gain new knowledge will still be deter-mined in most cases by the degree of resolution we can achieve, and not by the ability to detect faint emissions, or the so-called sensitivity of such instruments. This is exactly analogous to the situation which existed in radio astronomy until recently, when techniques were exploited which provided the level of high resolution needed to gain the full benefits offered by good sensitivity already available. Infra-red, optical and ultra-violet astronomy could well follow the same course.

Some of the most tantalizing and most rewarding potential discov-eries call for far better resolutions than 50 milliarcseconds, resolu-tions which could be obtained by the determined application of a mostly existent space technology. Here is a catalog of some of them.

Planetary System Formation

The process of star formation is obviously one of great importance. Understanding it would allow us to know just where the stars come from, and why their numbers in various sizes are as they are. More than that, it would tell us under what circumstances planetary systems are formed, what their numbers are, and where we would best look for them. We would learn what the distribution of planetary sizes in systems should be. For example, we would learn whether to expect systems in which most or all the planets are giant gas planets like Jupiter, or whether there should be systems with several or many planets like the earth. We would learn what starting conditions lead to what is probably a great variety in planetary systems.

We have recently made provocative observations which tell us that the answers are out there to be found. Observations with the IRAS, and later from the ground, showed that there are large numbers of nearby stars which are accompanied by disks of dust, in some of which the total amount of dust is enough to make hundreds of earths. It has long been thought that such a dust disk was the progenitor of the present solar system; indeed, we see the remains of our dust disk in the layer of dust known as the zodiacal light. In these other stars we see dust disks far greater in size and mass than the present zodiacal cloud of the Solar System. Are there in fact planets in these extrasolar dust disks, or agglomerations of dust which will become planets?

To distinguish planets from the bright light of the central star of a system is a formidable task. Even a large planet, similar to Jupiter, will be something like a billion times fainter than the star at optical wavelengths, and a million times fainter at infra-red wavelengths. The weak light scattered from the stellar image by the laws of optical diffraction and imperfections in the telescope may well overwhelm the faint emission from the planet. Therefore, to detect planets, an optical system of extremely high quality and resolution is necessary.

Studies have shown that a resolution many times better than the separation of the planet from the star is required. If we wish to be able to study even as few as several hundreds of stars, say to a distance of twenty parsecs (about 60 light years), we will need a telescope with a resolution of roughly 5 milliarcseconds or better. From the laws of optics, which can not be circumvented by any technical brilliance, the minimum dimension of the light collector must be about 100 feet for optical wavelengths, and perhaps 1000 feet for infra-red wavelengths. These are astonishing numbers, but they are specifications which can be met in space.

Astronomers have recently been enthused by the direct observation of the act of new stars being formed. Clouds of gas and dust, rich in molecules, have been seen collapsing under their own internal gravitation to become new stars and possibly planetary systems. We see that these clouds of gas and dust are actually spinning, as we have long expected. A new star can not form from such a spinning cloud unless the rotational energy in the cloud is ejected, probably through the formation of a spinning disk of dust, a progenitor of planets, just as observed.

A remarkable feature of many of these systems is that they eject two streams of material, presumably along the axis of the spinning, collapsing, system. These "bipolar jets" have been found associated with a wide variety of objects, ranging from protostellar objects, through exotic stars such as the one known as SS 433, to the cores of active galaxies from which the jets may extend into space many galactic diameters. It would appear that the same mechanism is at work in all of these objects to produce these jets, but we cannot be sure.

Clearly, we would like to see in detail the processes taking place in these places of star formation, in these sources of the jets. But the nearest of these objects is many hundreds of parsecs away. To see in detail the processes occurring in these early stages of star formation, we will need resolutions of the order of one milliarcsecond or better. This calls for optical systems whose maximum dimensions are of the order of 500 feet for optical wavelengths, and perhaps even a mile for infra-red wavelengths.

Quasars

The most brilliant, energetic phenomenon in the universe is the quasar. Visible from the farthest edges of the universe, and back in time to the period when galaxies were just forming, these objects appear from great distance (there are no close ones) as "points" of light. Yet we know that they are actually regions, probably about the size of the Solar System or smaller, in the center of galaxies. In these regions an incredible amount of energy is being released, an amount of energy which can be accounted for only through the annihilation—the total conversion of all mass into energy—of millions upon millions of stars.

They flicker on time scales of days or less. To see these regions in detail, at their minimum distances of many millions of light years, requires resolutions of the order of millionths of a second of arc. This requires, for optical wavelengths, observing instruments with dimensions of the order of a few miles.

From the core of many of these objects, enormous jets of material are ejected, material which has been accelerated to very nearly the speed of light. These clouds of particles move outwards into space as far as several galactic diameters, where the enegetic nuclear particles in them radiate powerful radio emissions.

What could these objects be? The most plausible theory is that they are basically very massive black holes, perhaps of a mass a billion times that of the sun. If this supposition is correct, it implies that stars and interstellar materials are being drawn into the black holes by their intense gravitation fields. As this occurs, this galatic material is accelerated to speeds approaching the velocity of light. When portions of it collide with one another, the enormous energy of their motion, given them by the gravitational field of the black hole, is converted to heat. The greatly heated material produces the intense radiation of the quasar.

To observe one of these regions in some detail would be a *piéce de résistance* for astronomy, a great leap forward in our ability to understand the nature of matter and its evolution. There appears to be one so close that it would offer an opportunity to observe the processes occurring within it in detail. It is the nucleus of our own galaxy. At present we do not see it clearly. Its optical light is obscured by extremely opaque dust clouds which lie between us and the center of the galaxy. But its optical and infra-red radiation does penetrate the clouds, and we have seen this object vaguely, as well as signs that a great deal of energy is being released from it. To observe it with resolutions equal to, say the dimension of the earth's orbit, which is highly desirable, would require instruments with dimensions of the order of six miles at infra-red wavelengths. This is a sizeable structure, of course, but would appear to be within the grasp of contemporary space technology. At radio wavelengths, the required instrument dimension is of the order of the diameter of the earth. It has already been demonstrated that this dimension can be synthesized by recording the radio emission received by radio telescopes in orbit, and subsequently collating this data in computers to duplicate the performance of a telescope whose size is about that of the orbits of the orbiting telescope or telescopes.

Steps to achieve this latter goal are already being taken. There is an international project called Quasat which is aimed at placing radio telescopes in orbit to achieve the required performance. This technique, called "Very Long Baseline Interferometry" is highly developed in many places now. In addition to a European project, the Soviet Union has for years been planning a series of radio telescopes in space, the telescopes to be placed at ever increasing distances from the earth, even as far as the moon. This system could achieve very striking resolutions within the galaxy and with distant galaxies and quasars. The United States, the original inventor of the technique and still by far the most expert practitioner of it, has a large ground-based system under construction, the Very Long Baseline Array, which will stretch from Hawaii to St. Croix. But we have been timid when it comes to the formidable space version of the system. Such a system appears on NASA's potential project lists, but with very low priority. Only limited studies are being made in support of the launch of an operational system.

Space Arrays

Beyond these radio systems, the observational opportunities beyond the Solar System resound with another leitmotiv. It is the need for telescope systems of hugh dimensions at ultra-violet, optical, and infra-red wavelengths. Dimensions of the order of 300 feet to several miles are called for. It is important to note that we understand that we do not need complete mirrors or lenses of this size. As has been amply demonstrated by the radio astronomers, it is enough to have only a few small mirrors which move around, in effect, over the area which would be covered by a huge mirror of the dimensions considered here. If an object is observed with such a system of mirrors, and the data subjected to the proper computer analysis, in the method called "aperture synthesis," the performance of a complete, large mirror can be synthesized. This is the *modus operandi* of a large number of instruments, including the wonderfully successful Very Large Array in New Mexico, and the various Very Long Baseline Interferometer systems which are in use.

What is needed here, then, is at least two small energy collectors, for example small mirrors perhaps a meter each in diameter. They must be held in a specific pattern in relation to each other, at separations of hundreds of feet up to miles, with extremely great precision. They must individually act as small telescopes to collect radiation from an object or region of interest. And the energy collected must be brought to a central collecting point with very precise controls. Two movable small telescopes at the ends of a long boom in space is a model of what is needed in the most minimal version of such systems.

We know from theory and experience that the use of more than two collectors offers considerable advantages. Indeed, the quality of the images produced by any such system improves roughly with the arithmetical square of the number of energy collectors in the system. The Very Large Array, for example, has twenty-seven individual telescopes in order to exploit this aspect of the technique. Thus, a large number of remarkable observing opportunities awaits the construction in space of systems with, say, ten or more telescopes mounted on long structures extending hundreds to many thousands of feet. Even larger numbers of energy collecting telescopes would be desirable. It also would expedite the acquisition of information, and allow clearer images to be constructed, if the systems had more than one arm. The land-based Very Large Array, for example, has three arms arranged in the shape of a "Y." Another much used geometry is two arms in the shape of a cross or a "T."

Although it is possible that a telescope array could be launched as a single payload, and this possibility needs to be explored, it would seem more likely that at least eventually the placement of such multi-telescope systems will require the construction in space of large structures, using components carried by a number of separate missions. It is likely to require human intervention to construct them, to put them in working order, and to maintain them with all their sophistication. This might be an ideal set of missions for a space station, although it will take detailed studies to show that this is the cost-effective way to achieve such systems.

As noted, systems intended to operate this way are being attempted on the earth's surface. It remains to be seen whether the deleterious effects of the turbulence in the earth's atmosphere, especially damaging to the efficiency of these systems, can be overcome. In any case, the clarity of images, the wavelengths available, and the range of brightnesses which can be measured with such systems will be strikingly greater in space.

Thus, it would appear that the astronomical conquest of space over the foreseeable future needs a series of single instruments like the "Great Observatories": the Hubble Space Telescope, the Shuttle Infra-Red Telescope Facility, the large X-ray and Gamma-Ray telescopes, and telescopes like the radio telescopes of Quasat. These may be launched by expendable launch vehicles, or, apparently at greater overall cost, by the space shuttle. In addition, we very much need the large aperture synthesis systems described above. Only the latter can reveal to us the true nature of many of the most fascinating and important phenomena in the universe. They are prime candidates for construction by the crews of the space station. Considering their complexity, they will almost of necessity need to be frequently altered, adjusted, and repaired. This would again seem to call for them to be associated with a space station.

Both expendable launch vehicles and a well-functioning space station seem to be very important to the future of space astronomy.

Gravitational Lenses

What of the more distant future? Surely it will be beneficial to build ever larger single instruments and aperture synthesis systems for space astronomy. In astronomy, bigger has always been better. No matter how powerful the telescope, there is always more detail to be seen, fainter objects to be detected, higher quality spectra to be obtained; goals such as these can only be achieved by ever larger instruments. The demands such instruments will make on our optical and instrument manufacturing facilities and our launch systems give one pause. But then the systems we readily build today seemed quite impossible only thirty years ago.

But the development of ever bigger versions of our current instruments is not the only path of evolution open to us. There is another quite fascinating path of instrument development which we know to be there, but which we have sampled in only the most primitive way. Unthinkable as it appears at first glance, this is to use our sun as the lens of a gargantuan telescope.

How does this work? The theory of relativity states that every mass in the universe "bends" space around it. The practical effect of this is that, with a star like the sun, light passing close to the sun does not travel a straight-line path, but follows a path which bends ever so slightly inward as the light passes the sun, so that the ray path eventually crosses a line from the source of the light through the sun. The typical bending of the ray path is about one arc second. One of the first triumphs of the theory of relativity was the actual detection of this light-bending at the edge of the sun, a phenomenon observed first at the time of a solar eclipse early in this century. The observation of this effect is still one of the prime tests of the correctness of the theory of relativity.

Now, imagine a large number of rays of light from a distant star passing close to the sun. All of these rays will be bent toward the imaginary line which comes from the star, passes through the sun, and goes on into space on the other side. The amount of bending depends on how close the ray passes to the sun; the closer it passes, the more a ray is bent. This means that all the rays which pass the sun at the same distance from its center, or those which pass through the same circle centered on the sun, will be bent the same amount. The "magic" in this is that at some substantial distance from the sun, all the rays will cross the imaginary line through the center of the sun. Such a phenomenon is identical to the achievement of an ordinary glass lens. Indeed, we describe this lensing behavior by masses as a "gravitational lens." All of the light from the rays passing through a circle around the sun are brought to a single point. They are focused!

An image of the distant star or other source of radiation is created at this focus. If there is an energy collector there, the equivalent of the detectors we put at the foci of ordinary telescopes, we can capture this image. We have constructed a telescope.

And what a telescope it is! The light comes to us through a ring around the sun, and thus through a "lens" whose diameter is greater than the diameter of the sun. This "lens" is very much like an ordinary lens without a center portion. The resultant resolution in the image is almost unbelievable. It depends on the wavelength we detect, with the exact behavior at different wavelengths yet to be calculated and tested. However, at radio wavelengths the resolution is something like one ten millionth of a second of arc—truly remarkable, but not surprising considering the size of the lens. To give a feel for the power of this resolution, it is equivalent to being able to resolve details only a hundred miles or so in size on objects—stars or planets—at great distances in our galaxy. The ability to use the gravitational lensing effect would allow us not only to detect planets many hundreds of light years from the earth, but also to see details on them, such as the existence of oceans and the rough shape of continents.

In addition to resolution, the light collecting ability of a solar or stellar gravitational lens is equally impressive. The effective collecting area can be literally hundreds of square miles, equivalent to hundreds to thousands of times the energy collecting area of the largest radio telescopes we have ever built or are even contemplating. This collecting area would allow us to detect even very weak intelligent radio transmissions from all the way across the Milky Way galaxy. One wonders if our television transmissions and a host of weaker signals might be regularly observed by creatures of distant worlds in our galaxy who use their stars as gravitational lenses—or perhaps are already being observed at one of the three hundred or so stars which are already being bombarded with our television transmissions.

We have observed the gravitational lensing of distant objects in about a half a dozen cases now. In each case so far, we have observed images of a distant quasar, with the image being created by the lensing effect of an entire galaxy along the line of sight between the quasar and us.

What about the sun and other stars? The use of a star as a gravitational lens creates a serious but not preposterous challenge. Since the maximum bending of light waves caused by stars like the sun is about one second of arc, as noted, the closest point where the rays are focused is distant from the star by about six hundred times the distance from the earth to the sun. This is about twenty times farther than the distance to the outermost planet. Other rays are focused at greater distances, by the way. So to detect the brilliant and finely detailed images created by a star, the observer must be as far from the star as at least six hundred times the distance from the earth to the sun. Obviously, we have not tested the technique yet with our sun. However, it is interesting to note that other stars will create images for us, but these will be very difficult to detect. A curiosity of this phenomenon is that white dwarf and neutron stars should create detectable high-quality images of their own planetary systems, if such exist. We have not looked for such images—it is a very time consuming operation, since the observer must wait until a planet, the star, and the earth are all in line.

Placing spacecraft at this distance, and maneuvering them, is within the reach of our present capabilities, but would certainly stress them. We are to be encouraged by the fact that spacecraft to go to these enormous distances are already being studied—for example, the "TAU" (for "Thousand Astronomical Unit") mission being studied at the Jet Propulsion Laboratory. There is no doubt that, to carry out this observing task well, we will need far more sophisticated space systems than we now have.

However, it seems very likely that such systems are the ultimate, most powerful, means to study the universe. They are somewhere in our future. For now, we can only wonder about their use elsewhere in the universe. In how many places are such instruments being used, perhaps even to study in detail the planet earth and its intelligent creatures, the humans?

ABOUT THE AUTHORS

EDWIN A. DEAGLE, JR., is manager of business operations for the Space and Communications Group of Hughes Aircraft Company. A graduate of the U.S. Military Academy and Harvard University, he served as a U. S. Army officer for 12 years. He has been on the staff of the Congressional Budget Office, the Office of Management and Budget and the National Security Council, and was director for international relations at the Rockefeller Foundation prior to joining Hughes.

FRANK D. DRAKE is dean of natural sciences and professor of astronomy and astrophysics at the University of California at Santa Cruz. He served as director of the National Astronomy and Ionosphere Center from 1971 to 1981. A former chairman of the astronomy section of the American Association for the Advancement of Science, he worked with several major observatories and was a member of the Cornell University faculty over a period of 20 years. He is widely known for his belief that life exists elsewhere in the universe and is a leading authority on methods for detection of extraterrestrial intelligent signals.

STEVE FETTER is a research fellow at Harvard University's Center for Science and International Affairs. His work on antisatellite weapons was undertaken when he was a postdoctoral fellow at Lawrence Livermore National Laboratory. A graduate of M.I.T. and the University of California at Berkeley, he has done research on reactor safety, nuclear weapon effects, energy and environmental policy and nuclear test ban issues.

MARCIA S. SMITH serves as a specialist in aerospace policy in the science policy research division of the Congressional Research Service. She was executive director of the National Commission on Space,

81

created by Congress and appointed by President Ronald Reagan, whose purpose was to formulate a new long-term civilian space agenda for the United States. The Commission's Report, *Pioneering the Space Frontier*, was released in mid-1986. Ms. Smith has been president of the American Astronautical Society and of Women in Aerospace and is a member of several other professional societies.